A *NEW YORK TIMES* NOTABLE BOOK OF THE YEAR

AN NPR BEST BOOK OF THE YEAR

A PASTE BEST NONFICTION BOOK OF THE YEAR

AN AMAZON BEST NONFICTION BOOK OF THE YEAR SO FAR

A JANET MASLIN MUST-READ BOOK OF THE SUMMER

AN *OUTSIDE* MAGAZINE BEST BOOK OF THE SUMMER

"A sea disaster tale unlike any other."
—*New York Times Book Review,* "New & Noteworthy"

"Intimate, eerie, and gripping."
—*Outside* magazine, "The Best Summer Books"

"A *Perfect Storm* for a new generation."
—BEN MEZRICH, author of *The Accidental Billionaires*

"A powerful reading experience.... As gripping as any fictional
thriller you're likely to find.... An exceptional work."
—The Maine Edge

"Harrowing, moving.... A taut adventure tale."
—*Boston Globe*

"An exhaustive account of what happens when tragedy
claims a vicious price for our progress and greed."
—Paste

"Chilling."
—*Star Tribune* (Minneapolis)

"The one account I've read that solves the
riddle of *El Faro* convincingly and thoroughly."
—ROBERT FRUMP, author of *Until the Sea Shall Free Them*

Praise for
INTO THE RAGING SEA

"A powerful and affecting story, beautifully handled by Slade, a journalist who clearly knows ships and the sea. . . . One can't help reading it, page after page, in disbelief."

—Douglas Preston, *New York Times Book Review*

"Damning. . . . A chilling account." —*Star Tribune* (Minneapolis)

"Powerful and gripping. . . . The depth of Slade's reporting is impressive . . . her storytelling ability even more so." —*Pennsylvania Gazette*

"Riveting." —Sam Sifton, *New York Times*, "Tastes of Summer"

"The depth of research and reporting, and Slade's skill at pacing and selecting the telling details produce a richly detailed narrative, tense and sad and true." —*Boston Globe*

"Rachel Slade mashes up *The Perfect Storm* with a suspenseful, page-turning thriller. This deserves a place on the bookshelf of modern maritime classics."

—Robert Frump, author of *Until the Sea Shall Free Them: Life, Death, and Survival in the Merchant Marine*

"A cautionary tale for leaders who think they have all the answers, for employees who choose not to speak up, and for organizations that rely on systems and processes that don't provide the information its people need to make the best decisions." —*Inc.*

"Immensely powerful. . . . Exerts a relentless grip that makes the book hard to put down, right to the closing pages. . . . Even for those who deal with these issues on a daily basis, this stirring book still shocks with the scale of the problems that it exposes. This is an important title, with lessons that extend far beyond the terrible tragedy that it describes." —*Nautilus Institute*

"A pulse-pounding, *Perfect Storm*–style tale. . . . A nerve-wracking, tension-filled narrative. . . . The author does solid work giving voice to the 33 mariners who lost their lives. The book serves as both a eulogy to them and a shout-out to the thousands of sailors who risk their lives every day to move goods around the world. A taut, chilling, and emotionally charged retelling of a doomed ship's final days."
—*Kirkus Reviews* (starred review)

"More than the story of how a ship was overcome by a storm, *Into the Raging Sea* is an allegory for what it means to be a part of the nation's largely invisible working and middle class." —Longreads

"A *Perfect Storm* for a new generation, *Into the Raging Sea* is a masterful page-turning account of the *El Faro*'s sinking, one that leaves you profoundly moved." —Ben Mezrich, author of *The Accidental Billionaires: The Founding of Facebook*

"In addition to a gripping narrative of a cargo ship's tragic voyage into the eye of a hurricane, Slade explains the fascinating world of commercial shipping and the essential—but often hidden—role it plays in our economy." —NPR, "Best Books of 2018"

"With skillful narrative prose and sensitivity, Slade takes readers on the final voyage of the *El Faro*. . . . Provides a haunting intimacy to this maritime disaster." —*Booklist*

"Well-crafted and gripping . . . Slade frames the tragedy with a meticulous review of all the ways in which it could have been avoided. . . . A painful and poignant narrative." —*Publishers Weekly* (starred review)

"Digs into the modern shipping industry and finds that countless bad decisions—many to please shareholders of the shipping company—placed ships and workers at great risk. [Slade] argues that with global warming, cutbacks to the National Hurricane Center, and lack of oversight of the aging merchant marine fleet, a disaster like *El Faro* could happen again."   —National Book Review, "5 Hot Books"

"Successfully, and very powerfully, navigates the difficult channel between insiders and outsiders. [Slade's] writing style will appeal to readers of nautical thrillers interspersing a fast-paced narrative of what was actually happening aboard the vessel as it collided with 120 mph eyewall of Hurricane Joaquin lurching through the Bahamas, with a rigorously researched backdrop covering commercial and regulatory issues. The secret sauce, if I had to point to one ingredient is the human side. Through interviews with families of the deceased crewmembers, and an insightful read of the transcript of what was actually happening onboard, taken from a voice recorder on the vessel's bridge, Slade is able to peer into the minds of those who perished."

—Seatrade Maritime News

"An extraordinary piece of reporting. Slade has accomplished what very few authors ever attempt: to explain the loss of a ship with no survivors. I tore through it like a novel but with the inside knowledge of how insulated the shipping industry is, how well it protects secrets, and the countless nets it deploys to entangle journalists. Slade pushes through the waves, heavy seas, and military court imbroglio in the same way *El Faro* faced hurricane Joaquin—dead on at full speed ahead."

—John Konrad, author of *Fire on the Horizon: The Untold Story of the Gulf Oil Disaster*

"It's almost unimaginable in this age of precise geography and instant communications that a massive, modern US ship should sail into the eye of a Caribbean hurricane and utterly disappear. With gripping prose and edge-of-the-seat momentum, Rachel Slade takes the reader aboard the final, fatal voyage of *El Faro*. *Into the Raging Sea* imparts a profound message about the power of nature and the fallibility of human judgment even in our digitized era."

—Peter Stark, author of *Astoria: Astor and Jefferson's Lost Pacific Empire—A Tale of Ambition and Survival on the Early American Frontier*

"A story as old as seafaring itself. Only, Slade isn't guessing here at the fate of the *El Faro*. This minute-by-minute account illustrates in chilling detail exactly what happens when the near-infinite might of the ocean plows broadside into the hubris of men."

—Brantley Hargrove, author of *The Man Who Caught the Storm: The Life of Legendary Tornado Chaser Tim Samaras*

"A gripping, moving account of a nautical tragedy, told with equal parts verve, gusto, and compassion. Don't miss it."

—Sarah Weinman, author of *The Real Lolita: The Kidnapping of Sally Horner and the Novel That Scandalized the World*

"Excellent and gripping. . . . Slade does an incisive and compelling job explaining what happened to the *El Faro*. . . . The lesson of *Into the Raging Sea* is that when the components of capitalism and global trade are not properly checked, regulated and restrained, and workers not cared for or respected, then lust for profit drives us all into the deep."

—*The Spectator* (UK)

INTO THE
RAGING SEA

INTO THE RAGING SEA

THIRTY-THREE MARINERS, ONE MEGASTORM,
AND THE SINKING OF *EL FARO*

RACHEL SLADE

ecco

An Imprint of HarperCollins Publishers

To the families and friends of those

lost on El Faro.

No one should have to endure such sorrow.

HarperCollins books may be purchased for educational, business, or sales promotional use. For information, please email the Special Markets Department at SPsales@harpercollins.com.

A hardcover edition of this book was published in 2018 by Ecco, an imprint of HarperCollins Publishers.

FIRST ECCO PAPERBACK EDITION PUBLISHED 2019.

Designed by Suet Yee Chong
Ship illustration by Michael Chan
Map by Mike Hall

Library of Congress Cataloging-in-Publication Data has been applied for.

ISBN 978-0-06-269987-9

19 20 21 22 23 LSC 10 9 8 7 6 5 4 3 2 1

There is nothing more enticing, disenchanting,

and enslaving than the life at sea.

—*LORD JIM*, JOSEPH CONRAD

CONTENTS

CAST OF CHARACTERS

DECK DEPARTMENT

Michael Davidson, 53, Captain

Steven Shultz, 54, Chief Mate, First Watch

Danielle Randolph, 34, Second Mate, Second Watch

Jeremie Riehm, 46, Third Mate, Third Watch

Frank Hamm, 49, Able Seaman, Helmsman for the First Watch

Jackie Jones, 38, Able Seaman, Helmsman for the Second Watch

Jack Jackson, 60, Able Seaman, Helmsman for the Third Watch

ENGINEERING

Richard Pusatere, 34, Chief Engineer

Michael Holland, 25, Third Assistant Engineer

LaShawn Rivera, 32, Chief Cook

James Porter, 40, Utility Person

Jeff Mathias, 42, Riding Crew Supervisor

***EL FARO* FAMILY AND FRIENDS**

Laurie Bobillot, mother of Danielle Randolph

Jill Jackson-d'Entremont, sister of Jack Jackson

Robert Green, stepfather of LaShawn Rivera

Rochelle Hamm, wife of Frank Hamm
Glen Jackson, brother of Jack Jackson
Jenn Mathias, wife of Jeff Mathias
Marlena Porter, wife of James Porter
Frank Pusatere, father of Richard Pusatere
Deb Roberts, mother of Michael Holland
Korinn Scattoloni, friend of Danielle Randolph

OFF-DUTY *EL FARO* OFFICERS

Charlie Baird, off-duty Second Mate
Jim Robinson, off-duty Chief Engineer

OTHER TOTE SHIP'S OFFICERS

Ray Thompson, *Isla Bella* Master, former Chief Mate of *El Faro*
Earl Loftfield, *El Yunque* Master after the loss of *El Faro*
Kevin Stith, *El Yunque* Master during accident voyage

JACKSONVILLE PILOT

Eric Bryson, River Pilot with St. Johns Bar Pilot Association

RETIRED MARINERS

Eric Axelsson, retired TOTE Master, Davidson's relief on
    El Faro until August 2015
Paul Haley, retired TOTE Chief Mate
Jack Hearn, retired TOTE Master
John Loftus, retired Horizon Shipping Master
Pete Villacampa, retired TOTE Master
Bill Weisenborn, retired TOTE Port Captain

SUN SHIPBUILDING AND DRY DOCK

John Glanfield, retired Shipbuilder
Eugene Schorsch, retired Naval Architect

NATIONAL WEATHER SERVICE

James Franklin, Director of the National Hurricane Center, Miami

US COAST GUARD DC HEADQUARTERS

Rear Admiral Paul Thomas, Assistant Commandant for
Prevention Policy

Captain Jason Neubauer, Head of USCG *El Faro* investigation

Commander Michael Odom, USCG Traveling Ship Inspector,
former Rescue Swimmer

Commander Charlotte Pittman, Deputy Chief, USCG Office of
Public Affairs, former helicopter pilot

Keith Fawcett, Marine Board Investigator

US COAST GUARD SEARCH AND RESCUE

Captain Rich Lorenzen, Commanding Officer, Air Station
Clearwater

Commander Scott Phy, Operations Officer, Air Station
Clearwater

Lieutenant Dave McCarthy, MH-60T Pilot, Aircraft
Commander for *Minouche* rescue

Aviation Survival Technician 1st Class Ben Cournia, Rescue
Swimmer during *Minouche* rescue

Lieutenant John "Rick" Post, MH-60T Pilot, Copilot for
Minouche rescue

Aviation Maintenance Technician 2nd Class Joshua Andrews,
Flight Mechanic during *Minouche* rescue

Lieutenant Commander Jeff Hustace, HC-130 Pilot, Aircraft
Commander for *El Faro* search

Captain Todd Coggeshall, Chief of Incident Management, 7th
Coast Guard District Command Center, Miami

Operations Specialist 2nd Class Matthew Chancery, Search
  and Rescue Mission Coordinator, 7th Coast Guard District
  Command Center, Miami

NATIONAL TRANSPORTATION SAFETY BOARD
Tom Roth-Roffy, NTSB Chief Investigator
Eric Stolzenberg, NTSB Nautical Architecture Group
Doug Mansell, NTSB Engineer
Mike Kucharski, NTSB Investigator

TOTE EXECUTIVES
Peter Keller, Executive Vice President, TOTE, Inc.
Phil Greene, President, TOTE Services
Phil Morrell, VP Marine Operations, TOTE Maritime
Tim Nolan, President, TOTE Maritime Puerto Rico
John Lawrence, Designated Person Ashore and Manager of
  Safety and Operations, TOTE Services
Jim Fisker-Andersen, Director of Ship Management, TOTE
  Services

CHAIN OF COMMAND
ABOARD *EL FARO*

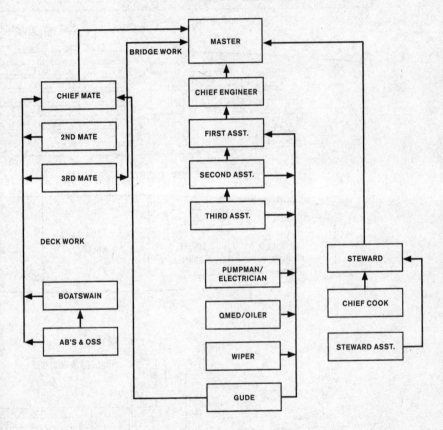

EL FARO PLANS AND SECTION

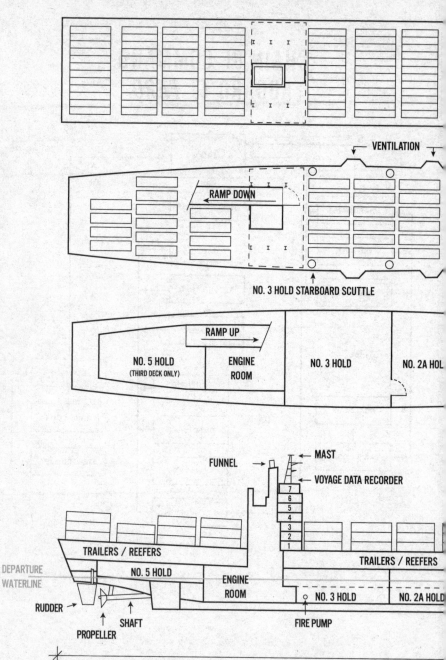

VENTILATION

RAMP DOWN

NO. 3 HOLD STARBOARD SCUTTLE

RAMP UP

NO. 5 HOLD
(THIRD DECK ONLY)

ENGINE ROOM

NO. 3 HOLD

NO. 2A HOL

FUNNEL

MAST

VOYAGE DATA RECORDER

7
6
5
4
3
2
1

TRAILERS / REEFERS

DEPARTURE WATERLINE

NO. 5 HOLD

TRAILERS / REEFERS

ENGINE ROOM

RUDDER

PROPELLER

SHAFT

NO. 3 HOLD

NO. 2A HOLD

FIRE PUMP

Length Between Perpendiculars 733 FEET 9 INCHES

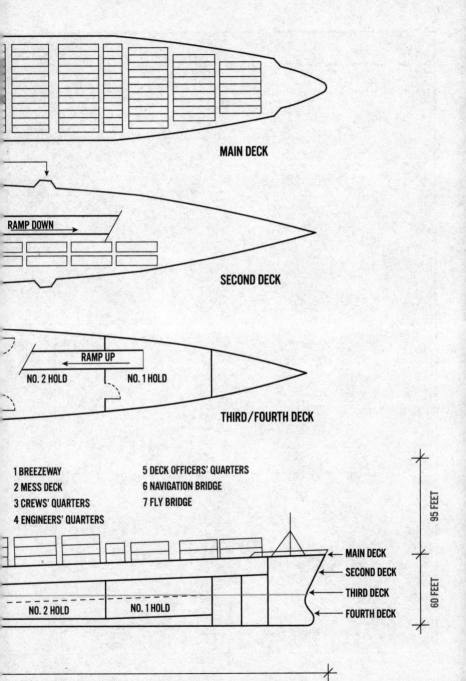

MAIN DECK

RAMP DOWN →

SECOND DECK

← RAMP UP

NO. 2 HOLD   NO. 1 HOLD

THIRD/FOURTH DECK

1 BREEZEWAY   5 DECK OFFICERS' QUARTERS
2 MESS DECK   6 NAVIGATION BRIDGE
3 CREWS' QUARTERS   7 FLY BRIDGE
4 ENGINEERS' QUARTERS

NO. 2 HOLD   NO. 1 HOLD

← MAIN DECK
← SECOND DECK
← THIRD DECK
← FOURTH DECK

95 FEET

60 FEET

Overall Length 790 FEET 9 INCHES

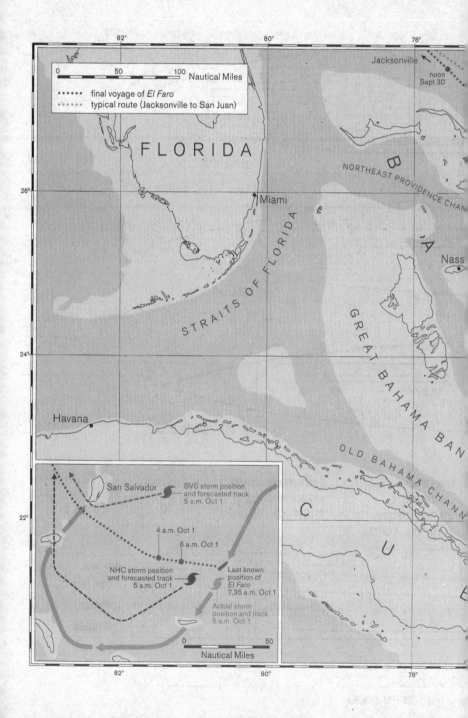

0   50   100   Nautical Miles

••••••   final voyage of *El Faro*
'•'•'•'•   typical route (Jacksonville to San Juan)

Jacksonville

noon
Sept 30

F L O R I D A

NORTHEAST PROVIDENCE CHAN

B

A

26°

Miami

Nass

S
T
R
A
I
T
S
  O
F
  F
L
O
R
I
D
A

24°

G
R
E
A
T
  B
A
H
A
M
A
  B
A
N

Havana

OLD BAHAMA CHANN

San Salvador

BVS storm position
and forecasted track
5 a.m. Oct 1

4 a.m. Oct 1

6 a.m. Oct 1

22°

NHC storm position
and forecasted track
5 a.m. Oct 1

Last known
position of
El Faro
7.35 a.m. Oct 1

Actual storm
position and track
5 a.m. Oct 1

C

U

C

0   50

Nautical Miles

82°   80°   78°

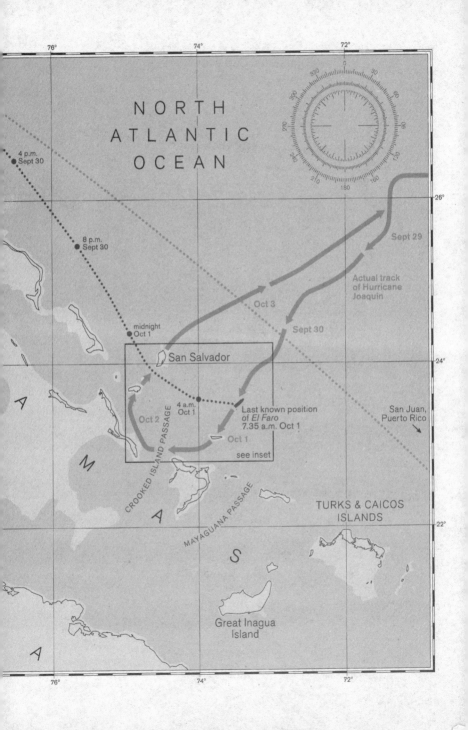

NORTH ATLANTIC OCEAN

4 p.m. Sept 30

8 p.m. Sept 30

midnight Oct 1

San Salvador

4 a.m. Oct 1

Last known position of *El Faro* 7.35 a.m. Oct 1

see inset

Oct 2

Oct 1

Oct 3

Sept 30

Sept 29

Actual track of Hurricane Joaquin

San Juan, Puerto Rico

CROOKED ISLAND PASSAGE

MAYAGUANA PASSAGE

TURKS & CAICOS ISLANDS

Great Inagua Island

A M A S

A

A NOTE ON THE TEXT

Six microphones installed in the ceiling of *El Faro*'s navigation bridge recorded twenty-six hours of conversation leading up to the sinking. This audio was captured on a microchip by an onboard Voyage Data Recorder—the ship's black box. All the dialogue in this book aboard *El Faro* during her final voyage was taken from a transcription of this audio.

PART
ONE

THE CLOCK IS TICKING

The satellite call came into the emergency center at 7:08 on the morning of October 1, 2015.

OPERATOR: *Okay, sir.*

CALLER: Are you connecting me through to a QI [Qualified Individual]?

OPERATOR: That's what I'm getting ready now. We're seeing who is on call and I'm going to get you right to them. Give me one second, sir. I'm going to put you on a quick hold. So one moment, please. Okay, sir. I just need your name please.

CALLER: Yes, ma'am. My name is Michael Davidson. Michael C. Davidson.

OPERATOR: Your rank?

CALLER: Ship's master.

OPERATOR: Okay. Thank you. Ship's name?

CALLER: *El Faro.*

OPERATOR: Spell that E-L . . .

CALLER: Oh man, the clock is ticking. Can I please speak to a QI? *El Faro*: Echo, Lima, Space, Foxtrot, Alpha, Romeo, Oscar. *El Faro.*

OPERATOR: Okay, and in case I lose you, what is your phone number please?

CALLER: Phone number 870-773-206528.

OPERATOR: Got it. Again, I'm going to get you reached right now. One moment please.

CALLER: [*Aside.*] And Mate, what else to do you see down there? What else do you see?

OPERATOR: I'm going to connect you now okay.

OPERATOR 2: Hi, good morning. My name is Sherida. Just give me one moment. I'm going to try to connect you now. Okay, Mr. Davidson?

CALLER: Okay.

OPERATOR 2: Okay, one moment please. Thank you for waiting.

CALLER: Oh God.

OPERATOR 2: Just briefly what is your problem you're having?

CALLER: I have a marine emergency and I would like to speak to a QI. We had a hull breach, a scuttle blew open during a storm. We have water down in three-hold with a heavy list. We've lost the main propulsion unit, the engineers cannot get it going. Can I speak to a QI please?

OPERATOR 2: Yes, thank you so much, one moment.

Thirty-three minutes later, the American government's network of hydrophones in the Atlantic Ocean picked up an enormous thud just beyond Crooked Island in the Bahamas. It was a sound rarely heard out there in the deepest part of the sea where, for decades, the government had been recording an endless underwater symphony. Three miles down, they listened to the lonely cries of humpback whales, the eerie hum of earthquakes, and the whirr of submarine propellers. Just white noise, really. But that morning, something huge and audible hit the ocean floor with terrific force.

Based on the positions of the hydrophones, the people listening knew approximately where the object landed. They also knew the precise moment that it hit. But what was it?

That the Americans had been listening in on the ocean since the 1960s was no secret, at least not to mariners. Some older guys remembered laying down the cable decades ago to feed this equipment, which served as the country's first line of defense against submarine invasion or other nefarious activity on the high seas.

The precise locations within this network were considered classified, but one monitoring station, known as the Atlantic Undersea Test and Evaluation Center (AUTEC), occupies a piece of Andros Island in the Bahamas, just west of Nassau. The thud was notable enough that there was talk among a few members of the armed forces stationed there. That intel simmered among a handful of officers assigned to monitor maritime activity in the Caribbean.

When word got out that a large American container ship had vanished in Hurricane Joaquin somewhere east of the Bahamas, those stationed on Andros Island knew exactly what they'd heard. It was the sound of *El Faro* colliding with the ocean floor.

CHAPTER 2

BLOUNT ISLAND

From above, Jacksonville looks like a battleground between land and water where water is winning. Rivers, streams, and inlets branch like veins folding in on themselves, following their own secret logic; roads curve here and there, searching for a path from one scrap of land to the next. A series of bridges stitches northern Florida's tenuous coast together.

It was late in the afternoon on September 29, 2015, and Jacksonville's wide, open-for-business highways were choked with commuter traffic baking in the heat of another late September day. Down at the sprawling marine terminal on Blount Island, stevedores loaded *El Faro*, a 790-foot-long ship, with 25 million pounds of cargo: 391 containers, 238 refrigerated containers, 118 trailers, 149 cars, and enough fructose syrup to make more than one million two-liter bottles of soda.

In his comfortable home at Atlantic Beach, Eric Bryson waited for a call from his dispatcher. Eric was a river pilot with the St. Johns Bar Pilot Association. He made his living divining the secrets of Jacksonville's waterways to safely deliver tankers, cargo ships, and car carriers up the St. Johns River to the Port of Jacksonville or back out to sea. Federal maritime law requires that every deep-draft vessel hire a local guide like Eric to navigate ships through these waterways so that they don't collide with each other or the struts of bridges, or take one of the many tight turns too wide and ground on the shallow banks.

To earn his piloting job, Eric had to memorize the twists, hazards, and depths of the St. Johns River in exquisite detail—enough that he could draw a navigable map of it from memory. Though the pilot's test was only open to seasoned ship captains like him, just the top scorer was considered for the position. When he took the exam in 1991, Eric blew away the other twenty-six applicants.

Since it requires such highly specialized knowledge, piloting is one of the most stable positions in the maritime industry. Getting the St. Johns job meant that Eric could give up the seaman's life and settle down for the long haul with his wife, Mary. They could work together raising their two kids in the warmth of the Florida sun, and he'd never be more than seven miles out to sea. It was a radically different life from the typical mariner who spends at least ten weeks straight on a ship, often out of communication, far from home.

Eric thinks of his job as a craft. "Any idiot can color inside the lines," he says of finessing a ship safely in and out of port. "The art of it is coloring outside the lines safely." Sturdily built, just shy of six feet, bald and bearded, with a face like a benevolent bulldog, Eric is the embodiment of male competence. Ponderous by nature, he does not take anything lightly. Some mariners who've worked with him call Eric "the Priest." All this, combined with his detached Yankee demeanor, puts ship captains at ease when Eric assumes command of their vessel.

The pilot dispatcher's call for SS *El Faro* came at six o'clock. The container ship was just about ready to leave port, but she was running an hour late. She'd been delayed because she'd been incorrectly loaded—someone had accidentally transposed a few numbers in the ship's loading software, causing the weight of the cargo to be unevenly distributed, which resulted in a noticeable list. The stevedores had had to scramble to shift cargo around to get her upright once again.

Eric knew *El Faro* and her two sister ships well—they'd been running twice weekly from Jacksonville to San Juan for nearly two decades, and he'd piloted them up and down St. Johns River dozens of times over the years. He knew many of the deck crew and officers, too, if not by name, at least by face. They were nearly all Americans. These days, that was notable. Most of the ships coming in and out of Jacksonville were registered in foreign countries and crewed by a mix of international laborers—predominantly Philippine.

By the time Eric got the call that hot Wednesday evening, his small travel bag was packed. He'd been following *El Faro*'s loading progress all day. He grabbed his navy-and-yellow pilot's jacket and walked through the kitchen door to one of his company's white Subarus, perpetually parked in the driveway.

As he drove to the terminal, a concrete expanse about the size of Central Park surrounded by a high chain-link fence, Eric cleared his head of the little things that might distract him. He needed intense focus to do his job right. Eric was acutely aware that in the world of big ships, complacency meant death. Now nearing sixty years old, he wanted to get out before his job killed him. Which it nearly did, twice.

To bring ships into Jacksonville, Eric had to board them at sea. He rode in his pilot boat seven miles out to the vessel, at first invisible on the horizon, just a blip on the radar, then looming large as he and his driver pull alongside. Riding up next to the massive, impermeable steel hulls never ceased to give Eric a thrill. If the winds and seas were high, the ship could trace out a big, lazy circle in the water to create a temporary lee for the small pilot boat.

Big ships rarely stop or drop anchor for pilots—they have so much momentum that they'd take miles to come to a halt, and even then, they'd drift with the winds and currents; instead, the pilot boat nestles up alongside like a pilot fish clinging to a whale shark, then tries to keep up with the ship, about 11 miles per hour. The two vessels aren't ever tethered together.

The small boat is outfitted with an external set of stairs leading to a platform, like a kitchen ladder that someone screwed onto its deck. Moored at the pilot association's office, the stepladder steps off into the void. But alongside a thousand-foot-long ship, it's the only way a pilot has any chance of climbing aboard.

Once the boat is riding alongside the ship, a small door in the enormous hull opens, a friendly face pops out from the steel wall, and a hand drops down a rope ladder as the pilot boat's driver tries to maintain the same speed and course as the ship. The pilot climbs his stepladder and then pauses, clinging to his platform, gauging the situation. The bottom rung of the huge ship's rope ladder dangles a few feet away from the pilot's shoes, but the two vessels are bobbing up and down at different rates. The pilot times his leap as the ship and pilot boat perform their version of synchronized swimming. He grabs hold of the rope ladder and scrambles up.

I once took a piloting trip with Eric. I thought I was mentally prepared for that moment when you had to take that leap of faith. But when confronted with all this—the flimsy rope ladder, the wall of black steel, the deep water rushing below with nothing to catch me if I missed—I panicked. Freud says that those who are afraid of heights don't trust their bodies not to jump. If my hands decided to let go, I would plummet between the two vessels and drown.

"Grab on! Climb up," Eric commanded. I reached out and gripped that rope ladder like my life depended on it, which it did. And then I scurried ten feet up the hull, ten steps straight up, holding my breath until I was safe on the deck, looking back down at Eric's head as he followed me scaling the side of the ship.

Pilots have missed that ladder. They've slipped and fallen into the sea where they were crushed between their boat and the ship. In October 2016, a pilot with more than a decade's worth of experience died this way on the river Thames.

Eric was once reaching for the ladder when his driver pulled away too soon, leaving him straddled between the two vessels. Eric was forced to make a split-second decision: jump for the ladder and risk missing it, or fall back into the pilot boat. He hurled himself toward the deck of his boat and saved his life but shattered his heel. He was out of work for months, and in immeasurable pain, but hated being on meds. His second accident was caused by his bag—caught on the pilot boat platform, it yanked him back, and when he fell onto the deck, he broke his back. He's never fully recovered from that. But he's alive.

Pilots board a berthed container ship like anyone else—up the gangway or loading ramp.

That's how Eric got onto *El Faro* on the evening of September 29.

He parked his car at Blount Island Marine Terminal at 7:30 and walked across the expanse of tarmac by the fading light to the ramp. As he approached *El Faro*, Eric regarded her dark blue steel hull. Like all massive ships, *El Faro* was a paean to modern engineering—a floating island capable of moving the weight of the Eiffel Tower through the ocean at 25 miles per hour. She was as long as the Golden Gate Bridge is high; it would take more than four minutes for a person to walk from bow to stern. Designed in the late 1960s, she was built for speed at a time when fuel was cheap and fast shipping fetched premium pricing. To reduce drag, she had a narrow beam that tapered steeply and evenly to her sharp keel.

These days, the profit is in cargo. The more you take, the more money you make. So modern ships are bulkier and slower; mega-ships, like the 1,305-foot-long *Benjamin Franklin*, are fat all the way to their little nub of a keel and can carry up to eighteen thousand containers. (If you lined them up, they'd reach from Manhattan to Trenton, New Jersey.) These ships made *El Faro* look like a toy.

Just about loaded up, *El Faro* sat low in the water, but Eric could still see her load line marks etched into her hull indicating she was safe to sail. Thick lines kept her tied to the dock as trucks and cars sped up her wide loading ramp, located at midships, in a fog of gas and diesel exhaust.

Twelve stories above Eric, a crane operator labored in a glass-bottomed cab hanging from a track, hauling the final containers onto *El Faro*'s main deck.

A working crane is like a massive version of the classic claw game, and watching it load ships is mesmerizing. On the dock below the crane, a longshoreman steers a flatbed truck into position next to the ship, then releases the clamps holding the container to the truck's chassis. A typical container is forty feet long and can be as heavy as thirteen cars. The crane operator uses cables and winches to lower the "spreader" over the steel box—a steel frame that aligns with the container's top four corners—and locks it in place like a tight-fitting lid. The spreader grips the container as the operator artfully swings it above *El Faro*'s deck, accommodating for the box's languid movement, and just as momentum begins to pull it backward, he lowers it into its designated spot, using gravity's pull to level it into place. Stevedores lash each box to the ship's deck with steel fittings, tightened with turnbuckles. Stacks of containers are locked to one another like Legos.

The frenzied choreography works most of the time but occasionally things happen—a container slips from the spreader's grip and crashes back on the flatbed, or it lands misaligned and crushes a longshoreman's hand as he's leveling it into place, or a stevedore gets hit by one of the huge trailers being driven off the ship in a great rush.

The huge loading cranes themselves also pose danger. They glide on two tracks that run parallel to the dock along its entire length, allowing the cranes to move from bow to stern and from ship to ship, stacking boxes. In 2008, a squall sent one 950-ton Jacksonville crane careening down the length of two football fields into the next, causing the steel constructs to twist and crumple into a heap of junk. It looked like spaghetti. It's a rare event—the cranes come equipped with multiple brakes—but shows the kind of power packed in gale-force winds.

Out of the din, Eric saw Jack Jackson coming down the ramp to greet him. Jack was the helmsman on the 8:00 to midnight watch that night; he'd stand at the ship's wheel on the navigation bridge and take steering commands from Eric. When they reached the sea buoy, Eric would disembark, leaving the captain in charge.

Sixty years old, sun-bleached yellow hair fading to white, ruddy face with a devilish grin, Jack had a wry sense of humor that could catch you off guard. He was the middle brother in a family of self-ascribed air force brats, mostly raised in New Orleans. Dad would show the kids pamphlets of the places they were moving next. Delaware! Beaches! Jack's older sister, Jill, moved frequently; she was a restless soul. Younger brother, Glen, with his ginger hair and blue eyes, could've been Jack's twin, but he was more a traveler of the mind.

When Jack and Glen were boys in New Orleans in the 1960s, there were four major shipping companies keeping the port busy. They used to ride out to the big bend in the river to watch the ships. Two sunburned boys dreaming on the banks of the Mississippi. The boats were always going somewhere, just like their family, always moving. That's where the fascination began, Glen says.

Jack signed up with the merchant marine during the 1970s to explore the world, make some money, live free. He enjoyed the easy sea life delivering grain to Africa or going back and forth to Europe. In those days, ships would spend days or weeks in a port loading and unloading so he could get off and poke around. Just one big adventure.

Jack had a magical aura about him; people were drawn to him. He once owned a motorcycle—a Norton. One day when Jack was stopped at a red light, a girl jumped on. She said she liked Nortons. They lived together for a year. That's how life was back then.

Jack was easygoing. He never pursued an officer's license or tried to hustle up the ranks; he preferred to avoid the messy politics aboard. He saw how cutthroat officers were; getting ahead wasn't worth the increase in pay and the hassle. Guys with higher ranks would sit in union halls for days waiting for work, but as a general able-bodied seaman, called an AB, Jack could always get work whenever the spirit moved him. Sailing, living free.

Decades later, Jack was still shipping out, and in a way, so was Eric. But Jack had fully embraced the seaman's life; he lived alone in Jacksonville and had never married. During shore leave, he carved sculptures out of wood and painted dreamy Edward Hopper–like scenes. He once told Eric that when he retired, he planned on throwing an oar over his shoulder and walking west until someone asked him what it was.

Eric was pleased to see his fellow traveler on duty that night. Guys like Jack made the heavy workload on these short, busy runs down to San Juan go a little smoother.

Working ships are an obstacle course of cargo, safety equipment, and machinery and as Jack led Eric from the ramp to the ship's house, he casually pointed out anything—raised thresholds, low headers—that Eric might trip over or bang his head on.

Though *El Faro* had an elevator, the captain on duty didn't want the crew using it. Someone could get stuck in there. The machinery was old and cranky, and he worried that it might stop working entirely because the shipping company had stopped paying to have it inspected. If it did quit with someone inside, the ship's engineers would have to spend their time fixing the lift instead of tending to the ship's ancient steam plant.

So Jack and Eric climbed seven flights of industrial stairs bathed in fluorescent light to get to the navigation bridge. Each floor in the ship's house served a function. Ascending, they passed the galley, the crews' quarters, engineers' quarters, deck officers' quarters—each door clearly labeled—to the final set of stairs, through the door to the bridge where the captain and the third mate were waiting for them.

Nearly ten stories above the waterline, the navigation bridge was a single room atop the house with seven windows framing the evening sky as the sun set over Jacksonville. It smelled like old solder and aging wiring, an olfactory blast from the 1970s, a reminder that the ship was forty years old—built before the internet, laptops, and cellphones—and nearly double the age of the average ship docked in America's ports.

Against the windows sat a file-cabinet-gray metal console as wide as the bridge, strewn with instruments. In the middle of the console, waist-high, was a small black steering wheel, like something you'd find on a Honda Civic. It was hard to believe that such an inconsequential thing controlled the massive rudder below. A couple of solid wood grab bars were attached to the console; even on this giant vessel, things could get rough. Tucked in the corner was the chartroom; a blackout curtain took the place of a door.

Captain Michael Davidson greeted Eric. Davidson was a Yankee like the pilot. He had signed on with the shipping company—TOTE—two years ago, and the pair had met here on the bridge a few times. Davidson stood a few inches shorter than Eric and wore his white-and-gray hair close cropped, military style. He had an athletic build, sharp nose, and Scottish eyes that sloped down in a friendly, generous way when he smiled, complementing his cleft chin. A slightly imperious manner betrayed his station as master of the ship. He shook Eric's hand firmly, calling him "Mr. Pilot."

"Hello, Cap'n," Eric said.

Eric opened the door to the bridge wing and stepped into the warm evening air, high above the river. Below him, a pair of pelicans flew over the stacks of containers. Eric had a portable GPS system that came with an antenna that worked best out here; he positioned it carefully on the small deck, sheltered from the wind. The antenna sent his precise location back inside to his iPad, which was loaded with a detailed chart of the St. Johns River, its depths, and live traffic information. The keel of *El Faro* cut a wake three stories below the waterline, a comfortable depth for the river's forty-foot dredged channel, but at various locations, tolerances were tight.

Eric's handheld software was more sophisticated than *El Faro*'s onboard technology. *El Faro* didn't have an electronic chart system, soon to be required of all international deep-draft vessels. Instead, her officers relied on their paper nautical charts published by NOAA (National Oceanic and Atmospheric Administration), and a crude alphanumerical GPS interface.

Eric always steered ships with his eyes first, technology second. He knew the river like he knew the streets of his own neighborhood, but he regularly referred to his iPad to check his work.

Jack Jackson handed the pilot a cup of coffee and took the wheel at eight o'clock as two tugs below slowly spun *El Faro* 180 degrees to point her bow toward the Atlantic Ocean, 10.6 miles down the winding river. When she was in position, the tugs pulled their lines and *El Faro* glided silently forward under her own power. They were underway.

Eric stood in the dusk at the windows of *El Faro*'s bridge sipping his brew, watching the waterway bend before him. Lights from the houses along the shore revealed the river's edge, but Eric closely monitored his iPad tracking the ship's speed and position and the river's depth. Stretched out before him was a field of red and blue cargo containers, loaded with every necessity for life in Puerto Rico, graying to black in the fading light.

At this time of night, there wasn't much traffic on the river. The navigation bridge was quiet, illuminated only by the glow of the ship's instruments as Jack stood at the steering wheel, watching for shrimp boats, pleasure boats, anything the huge ship could crush as it slowly made its way downriver.

In his resonant baritone, perfect for late-night jazz radio, Eric would occasionally call out a command, anticipating current changes ahead to keep the vessel on course. "Left 10."

"Left 10," Jack Jackson repeated from behind the wheel and turned accordingly, causing the ship's massive rudder far below to swing ten degrees to port. The heavily loaded *El Faro* heeled over slightly with each turn. She was tender, sensitive, slow to right herself, but not unpleasant. That's how this class of ship moved.

As usual, there was very little chatter among the people on the bridge, just another trip on a calm autumn evening down the coast of Florida. Mariners are comfortable standing in silence for hours at a time, staring out at the sea. It's part of the job. Jack and Davidson talked a little about the weather with Eric. Something was brewing out there, but *El Faro* was a fast ship. Davidson said that the storm would cut north and they would shoot down under it.

Eric could hear Second Mate Danielle Randolph on the two-way radio overseeing the men securing the deck and organizing lines at the stern of the ship. She'd spent the whole afternoon directing stevedores and checking cargo. Soon she'd head back to her room to catch a few more hours of sleep before coming up on the bridge for her midnight watch.

"Dead slow ahead," Eric said as they approached land's end. When they reached St. Johns Point, just past Mayport Naval base, the coastline peeled back as *El Faro* moved into open waters. So far above it all, Eric could feel the thrum of the twenty-five-foot propeller turned by the ship's steam turbine whenever they put on the rudder. By the time the ship reached the mouth of the river, stars were out. The vessel and her cargo were a black silhouette against the dusky sea and sky.

About seven miles out, just before ten o'clock, Eric retrieved his GPS antenna from the bridge wing, packed his bag, and shook the captain's hand. Chief Mate Steve Shultz walked him down to the main deck. Eric looked over the side and saw his pilot boat pulling up alongside. He gingerly climbed down the rope ladder, made it safely to his ride, gripped the handrail of his boat and looked up. From the top of the ladder, Shultz smiled and waved. The pilot boat pulled away from *El Faro* and headed back toward the lights of Jacksonville.

CHAPTER 3

TROPICAL STORM JOAQUIN

29.07°N -79.16°W

El Faro steamed south throughout the night hugging Florida's Atlantic coast.

Two hours before dawn on September 30, Captain Davidson and Chief Mate Shultz met on the ship's bridge. Since leaving Jacksonville the night before they'd traveled 147 nautical miles, putting them eighty miles east of Daytona Beach. Now they needed to make a decision: continue on their direct route to Puerto Rico or take a southerly detour along the west side of the Bahamas.

Above them, the night sky was clear. A handful of Tropical Cyclone Advisories had come in from the National Hurricane Center (NHC) overnight. Each time they did, a dot-matrix printer above the satellite-fed computer would automatically type out the message. The latest was advisory number 10:

```
TROPICAL STORM JOAQUIN FORECAST/ADVISORY NUM-
BER 10: 0900 UTC [5:00 a.m.] WED SEP 30 2015:
```

```
TROPICAL STORM CENTER LOCATED NEAR 25.4N 72.5W
AT 30/0900Z. PRESENT MOVEMENT TOWARD THE
WEST-SOUTHWEST AT 5 KT. MAX SUSTAINED WINDS
60 KT WITH GUSTS TO 75 KT.
```

In the silence, darkness, and coolness of the final hour before dawn, Davidson used his chief mate as a sounding board as he tried to determine whether Joaquin posed a threat.

Shultz was taller and softer than Davidson, and a year older. His thin brown mustache and sparse goatee obscured a smattering of acne scars. As chief mate, he was responsible for the loading and securing of cargo and overseeing the unlicensed crew. Although weather routing was the second mate's job, Shultz was flattered that Davidson had opted to consult with him that morning.

"This NHC report puts the storm further south than last night," Shultz observed, studying the newest message out of Miami.

Davidson didn't like Tropical Storm Joaquin's slow, lumbering movements. It was traveling at about four knots. And the newest NHC assessment of the storm—that it was heading southwest—contradicted the ship's forecasting software, which had the storm turning north. "Look, remember how we saw this storm out here the other day? It's just festering," he said to Shultz. "I'm anxious to see the newest BVS report."

El Faro was equipped with a third-party weather forecasting software package called Bon Voyage System (BVS). Its interface is lush, much more visually inviting than the all-caps text advisories coming from NOAA, which need to be plotted out by hand on the ship's paper charts.

On BVS, weather comes preplotted on a pastel-hued digital map. The tiny islands of the Bahamas and Turks and Caicos appear as beige strands in a light blue sea. Mariners can plug in their ship's course and fast-forward through time to watch how weather systems are predicted to behave as they sail to their destination. Click and it's tomorrow. Here's your ship (based on your projected speed and course) and here's the weather. Click again and it's two days from now. You've moved, the weather moved. Click, click, click. It's five days in the future, and there you are, safely in port. It all looks so clear, so real. Solid and dependable.

Nearly all people glean more information faster from data visualizations rather than alphanumeric code. That's just how our brains work. Davidson, whose elegant handwriting reveals a man with a strong aesthetic sense, preferred to get his weather from the highly graphic BVS software. It put all the information—ship and storm location, plus their projected courses over time—on a single chart. On the bridge, this information was recorded on separate paper maps, one for the ship's course, one for the storm, requiring a mental workout for an officer to precisely understand their relative positions. "I'm connecting now," Davidson told Shultz and retreated downstairs to his stateroom to download the 5:00 a.m. BVS update via satellite.

It took a minute for his desktop to get the information and process it, but once it did, a small red circle with two helical wings represented Joaquin. It hovered far east of the Bahamas. Around the storm was a series of wavy concentric circles—bands of color depicting predicted wind speed from dangerous orange to so-so yellow to benign light green, then darker cobalt, and finally, the neutral baby blue of a calm sea. A dark scarlet line showed Joaquin's predicted path.

The storm would move a little farther southwest, BVS told Davidson, then cut north toward South Carolina. El Faro could easily skirt below the system, as long as she kept up her speed. The captain clicked the forward arrow and saw the future: tomorrow, the ship would be halfway to San Juan and Joaquin would be nearing the US coast. As he clicked forward in time, El Faro moved along its expected route, and Joaquin inched farther and farther north, away from the islands. They'd be fine.

What Davidson didn't know was that due to a clerical error, the 5 a.m. BVS forecast he'd downloaded that morning was identical to the one sent six hours prior. Because BVS took several hours to process NHC data before issuing a report, that error meant the weather forecast Davidson was looking at was based on raw data nearly eighteen hours old. The report depicted Joaquin as a northbound tropical storm when, in fact, by dawn on September 30, most forecasters, including the NHC, recognized the system as a slow-moving, full-blown Category 3 hurricane that might not budge from its southwesterly track.

According to Davidson's BVS report, they'd see some weather—winds and waves, maybe twelve to fifteen feet—the aftershocks of the storm. He sent the update to the computer terminal on the bridge for the other officers to see. "This doesn't look too bad," Shultz said, examining the projection when the captain joined him. "The ship can handle it."

"We'll see what the schedule looks like," Davidson said, clicking through to Friday. He took the opportunity to lecture his new chief mate on the finer points of routing a ship. "Joaquin's gone a little south," he said. "This is why you watch the weather all the time. All the time." He cleared his throat then added, "Absolutely."

As dawn approached, *El Faro* was one hundred miles north of the Bahamas. If they maintained their current heading, they'd sail east of the island chain into the deepest Atlantic—a straight shot to Puerto Rico. But if Joaquin didn't turn north as predicted, they could get into trouble.

A land map shows the Bahamas as a series of spindly islands and cays running from the Florida peninsula to Cuba's eastern tip. A nautical chart or satellite photo tells a very different story: the Bahamas are, in fact, the highest ridges of two huge limestone masses built up over millennia by the creation and compression of coral reefs. Known as Little Bahama Bank and Great Bahama Bank, these plateaus were once dry land before sea levels rose following the last Ice Age, creating extremely shallow areas that block deep-draft vessels from accessing the Gulf of Mexico from the Atlantic.

On a calm day, the banks' waters are a hypnotic cerulean, easily distinguishable from the dark, deep waters around them. But on rough and windy days, when the water surface is shredded by winds, few landmarks will warn you that you're about to hit a submerged wall, until you do.

A handful of underwater canyons provide deep but narrow shipping highways. The Northwest Providence Channel, for one, runs east to west, separating Little Bahama Bank to the north from Great Bahama Bank to the south.

The Old Bahama Channel is another natural canyon—a fifteen-mile wide chasm between the Great Bahama Bank and Cuba. It's such a popular route that NOAA's nautical chart number 11013 shows it as a divided highway, complete with purple arrows reminding mariners that traffic here follows the right-hand rule. Tolerances in the channel are tight, though, bound by hidden seventeen-hundred-foot-high cliffs that plunge down into the abyss.

For centuries, mariners have threaded their way through these channels, sounds, and astonishingly deep trenches, losing countless ships when storms forced them onto the unforgiving shoals, leaving an estimated $400 million worth of Spanish plunder between Cuba and Florida. In August 2015, just a month before *El Faro*'s ill-fated voyage, treasure hunters announced the discovery of a Spanish galleon sunk by a hurricane in 1715 off Florida's coast, a find that yielded more than $1 million worth of gold coins.

After countless casualties, prudent mariners started favoring the Straits of Florida, which provide wide and deep passage from the southern tip of the Sunshine Sate along Key West, clear all the way to the northwest coast of Cuba, and into the Gulf of Mexico.

On the Puerto Rico run that morning, playing it safe meant following that same line: adjusting *El Faro*'s course 50 degrees south, *right now*, and taking the Straits down along the lee side of the island chain to the Old Bahama Channel. That would add 160 miles to the voyage, and at least five hours to their arrival time.

One month earlier, Second Mate Charlie Baird had convinced Davidson to take this same route to avoid Tropical Storm Erika, a move that shifted their ETA six hours later and burned 504 more gallons of fuel. Erika turned out to be a dud of a storm. It never developed into a hurricane and rumor has it that TOTE wasn't pleased with Davidson's decision to take the long way around. In hindsight, it seemed like an overcautious move that cut into the run's profit. Fuel costs $38,000 a day to keep *El Faro* running full speed ahead. Port labor is also pricey. Time and fuel—it was always a delicate balance.

There was pressure to save on both, and Davidson had blown it once. He wasn't going to blow it twice in the same season.

Davidson knew that early on Friday morning, two days from now, stevedores and hundreds of trucks would be lined up at San Juan Port waiting to unload *El Faro*'s cargo. By Friday afternoon, supermarket shelves across the island would be fully stocked with the goods she carried. By Monday, provisions would be low, and the trucks would line up again for *El Yunque*, *El Faro*'s sister ship, to bring another haul. TOTE's ships were Puerto Rico's lifeline. Any delay in arrival would set off a costly chain of events.

Davidson couldn't afford to screw up with TOTE again. In the cutthroat world of shipping, prudence didn't always pay. As a master, you had to be willing to push your luck. He'd lost a good job once before for playing it too safe when he ordered a tugboat assist after one of his ships developed steering problems.

Davidson decided to take his chances with Joaquin. Atlantic hurricanes always cut north eventually, and the ship was fast. They could easily outrun this one.

He and Shultz marked a point on their map east of the Bahamas and set a course nine degrees more southerly than the normal
route, a slight change which they hoped would keep them out of
trouble. "I think that's a good little plan, Chief Mate," Davidson
said after spending an hour hashing it out. "At least I think we
got a little distance from the center. We're gonna be further south
of the eye, about sixty miles south of it. It should be fine. We are
gonna be fine. Not should be—we are *gonna* be fine."

He knew they'd probably feel the aftereffects of the storm and
wanted the ship secured for heavy weather. "Take a hard look at
some of that cargo down there," he advised his chief mate. "Del-
egate the men to look at the lashings as you deem necessary."

Davidson's warning about the lashings made the chief mate
reconsider what he'd witnessed the day before when the stevedores
were loading the trailers and containers onto the ship. He didn't
think they'd secured them properly but didn't make a big deal out
of it. "The longshoreman was doing the lashing wrong and I was
trying to help," Shultz told Davidson. He was the new guy on the
ship and felt awkward criticizing their work.

"Go right to the foreman—cut out the middleman," Davidson
advised him. "I do it all the time."

"They just don't do the lashing the way it outta be done,"
Shultz repeated in frustration. It was his job to make sure loading
was done right, but it had all happened so fast, and he hadn't yet
established solid relationships with the workers he was overseeing.

Shifting cargo doesn't only cause damage, it can break loose
and kill a man. It can go right over the deck or wreak havoc in
the holds. It can set off a domino effect during heavy seas, throwing the ship over, causing a perilous list. Globally, ships lose an
average of about six hundred containers each year to storms, collisions, and groundings. Now that they were at sea, Shultz worried whether they had enough chains to double- or triple-lash if
they needed to. If he'd looked at the cargo-securing inventory,

he'd know that the vessel was short 170 lashing rods and missing more than a thousand required turnbuckles and twistlocks. Even if he had the equipment on hand, though, trying to lash at sea was miserable—it's nearly impossible to maneuver around a loaded ship like that to chain things down. The cargo is packed tight, and if something did come loose while you were trying to secure it, you could be crushed. They were too far gone. Hopefully, the chains would hold.

Shultz spent the next several minutes logging alternative routes and waypoints into the ship's GPS system. He wasn't completely sure he trusted Joaquin, and he wanted to give the other officers various options if they ran into trouble. The storm had proven an unpredictable and erratic foe; Shultz wanted to work out as many escape routes as he could.

Logging in all those new waypoints (which made a high-pitched beep with each entry) irritated Davidson.

"It's a good little diversion," he scoffed at Shultz. "Are you feelin' comfortable with that, Chief Mate?"

"Better. Yes, sir."

"You can't run every single weather pattern," Davidson barked. His chief mate was already rubbing him the wrong way, and they'd only been up for an hour.

Shultz didn't want to piss off Davidson. He was brand-new to the ship, new to the captain, and relatively new to the Atlantic. He'd come from shipping for years in the Pacific Northwest and was happy to be on the Puerto Rican run, at least temporarily, closer to Florida where his petite, native Brazilian wife and two teens lived. He was a dedicated husband who vigorously embraced his wife's Catholicism.

Nothing was more sacred aboard a ship than respecting the chain of command. Though he had as much experience as Davidson, Shultz knew to defer to his captain in all things to ingratiate himself to his superior. If he thought that sixty miles from the eye of a major storm didn't sound right, he kept it to himself. A challenge like that could be viewed as insubordination. Shultz had kids approaching college age. Like everyone else, he had expenses. He needed this job.

Outside in the wild Atlantic, however, there were ominous signs.

At 6:40 a.m., Davidson watched the rising sun set the eastern horizon aflame, reminding him of an old adage. "Look at that red sky over there," he said to Shultz. "Red in the morning, sailors take warning. That *is* bright."

For thousands of years, sailors have viewed crimson skies at dawn as a bad omen. Science backs this up: Red has the longest wavelength in the color spectrum—powerful enough to penetrate the dust and moisture kicked up by an atmospheric event, such as a major storm. Other colors in the rainbow get scattered by thick storm clouds, leaving the sky a blazing scarlet.

Dawn on September 30, 2015, was distorted by a terrific atmospheric event brewing dead ahead of *El Faro*.

CHAPTER 4

THIRD MATE JEREMIE RIEHM

28.42°N -78.47°W

Third Mate Jeremie Riehm arrived on the bridge shortly before his 8:00 a.m. shift and watched the tropical sun rise into an unctuous sky. Another steamy Caribbean day ahead. The air-conditioning had been cranking all night and now it was chilly on the bridge in spite of the stickiness of the morning outside.

The scorching sun would soon beat down on the metal roof of the wheelhouse, so Jeremie wore a T-shirt and shorts under a light jacket. His thick brown hair flopped into his eyes. He'd brusquely brush it up and back with his fingers, giving it a permanent flip. Forty-six years old, he looked like a rugged, slightly stockier Brad Pitt.

"You hear about the storm?" Chief Mate Shultz asked Jeremie.

"Yes, I'm aware of it," he answered. "Caught a little bit of it on the news. I heard it's gonna be a hurricane tomorrow or later today according to the Weather Channel."

Shultz took Jeremie over to the chart table to show the new course he and the captain—who was downstairs getting breakfast—had prepared earlier that day. The chief mate pointed to Joaquin's projected course, drawn in wax pencil on a clear plexiglass panel laid over the paper nautical chart. Tiny numbers scattered across the white sea indicated depths in fathoms; there was twelve thousand feet of ocean below them. The shallow Bahama banks blocking their escape route to the west appeared as a scorpion-shaped expanse of light blue. Sinuous purple lines represented the elaborate network of underwater communication cables that link the United States to the Caribbean islands, South America, and Europe—cords that could be cut by the keel of a ship if it dragged over them.

Joaquin had moved farther west and south, which meant that if they'd continued on their original course, they'd run straight into it, Shultz said. "I mean, our timing was perfect to reach the eye. Now worse comes to worse, we can duck in behind the islands."

"We're gonna get slammed tonight, though," Jeremie said grimly, studying the chart in front of him. A slight heading change wouldn't buy them much, he thought, and sixty miles from the eye was at least forty miles too close.

Jeremie looked up to nod at Jack Jackson, his helmsman on the 8:00 to noon watch, who'd just come up from breakfast. Jack stood by while the chief mate and third mate discussed the new plan. "It's been too quiet this season," Jack observed.

"Been that way last year and the year before," agreed Jeremie. Very quiet hurricane seasons. Good weather could breed complacency.

"Out in the Pacific, it's been one typhoon after the other, a daisy chain of 'em," Jack said. "One in Taiwan hit 180 miles per hour." He shook his head thinking about that kind of wind. It was enough to give an old mariner like himself a good scare.

Davidson came back up to the bridge to check in on his third mate. "So we got a little weather coming in," he told him. "I'm sure you heard," he continued. "Joaquin morphed its ugly head between the time we left and the morning when we woke up. Tough to plan when you don't know, but we made a little diversion here. We're gonna be further south of the eye. We'll be about sixty miles south of the eye. It should be fine. We are gonna be fine—not *should* be—we are *gonna* be fine."

The captain looked at his third mate intently, confirming that the officer understood the plan. Satisfied, he left the bridge to Jeremie and Jack who together watched the rising sun cook the already hot Caribbean waters into a soupy haze.

Jeremie was known for speaking his mind. Unlike the other officers aboard *El Faro*, he'd come up through the hawsepipe, meaning he hadn't graduated from one of the country's seven maritime academies. Instead, he'd signed on to the merchant marine as general crew when he was eighteen years old and worked his way up. During his downtime aboard the vessel, he'd studied the ship's books and manuals and often surprised his fellow officers with the depth of his knowledge. Most of them were older and had little interest in exploring all the bells and whistles offered by advancing technology. They could run a ship with or without it. But Jeremie liked figuring things out.

Once he had enough proficiency and experience, Jeremie screwed up his courage and took the third mate's test. The exam is a brutal, three-day trial of knowledge and nerves. It was especially hard for Jeremie, who got anxious at the thought of taking tests. After passing, he didn't want to go through that again, so he remained working as a third mate for the next decade. Sailing as a second mate would have meant more money and more authority on the bridge, but for him, it wasn't worth the stress. He lived modestly on shore, focused on family. He was married to an African American woman who ran a day-care center on the remote

Florida island where they lived with their two teenage children. When he was off duty, he helped her with the business and kept to himself.

A few years back, Jeremie joined his fellow ship's officers Captain Pete Villacampa, Chief Mate Paul Haley, and Second Mate Charlie Baird at their union's ultra-advanced simulation center in Miami. He was a generation younger than the other officers. For four days, they worked together on a computer-generated bridge, which rocked and rolled like a real ship, enough to make a person seasick, as the program generated complex maritime situations for them to work their way out of. Jeremie's understanding of the weather systems, radar, and loading software—things he'd taught himself—proved invaluable. The team earned one of the highest scores the instructor had ever recorded.

Three months after that simulation test, Villacampa and Haley were fired by the shipping company, along with two other seasoned officers, Captain Jack Hearn and Chief Mate Jimmy Armstrong. The official line was that they'd lost their jobs because illegal drugs had been found on one of TOTE's ships.

In July 2012, US customs agents in Jacksonville saw a suspicious shipping container coming off *El Faro*'s sister ship, *El Morro*. The box looked like it had been tampered with, maybe pried open after it had been sealed. Sure enough, a couple of the unlicensed crew—an ever-changing cast of characters hired through the union by the shipping company's crewing manager—had stashed forty-seven kilos of cocaine inside it, a $3.5 million haul bound for the US market. After the seamen were busted in Jacksonville, the ship's steward—a Puerto Rican named Danny, and another family member—were gunned down by a member of the drug cartel in a San Juan restaurant while Danny's vessel was docked nearby.

That shipping containers had been used to transport contraband surprised absolutely no one. Roberto Saviano's book on the Italian mafia, *Gomorrah*, offers a shocking example of the disturbing things people stuff in these nondescript steel boxes. In his book's opening scene, a crate being loaded onto a ship in Naples accidentally opens midair and dozens of human corpses pour out of it and onto the ship's deck. The dead immigrants had paid a lifetime's worth of wages to the Mafia to have their remains repatriated to China; they were unceremoniously scooped back in and shipped according to plan.

No port authority has the resources to monitor what's inside the hundreds of thousands of shipping containers crossing the oceans at any one time. Worldwide, approximately 1 percent of the boxes are actually opened and inspected. Drug dealers and arms brokers count on this fact to move their illicit wares around the globe; they build rare losses into their business plan. Because occasionally, someone gets caught.

Jack Hearn was captain of *El Morro* when the drugs were found, but how could he have known about them? Jacksonville Port was known as a major gateway for drugs traveling from South America to the United States, especially since Puerto Rico's economic collapse created a jobless, desperate population on the island. Of course, TOTE's ships occasionally carried contraband.

Hearn's job was to deliver cargo and keep the vessel and crew safe. He'd done just that for more than thirty years. In that time, he'd watched the profession go from sextant to satellite. And in that time, the role of the captain evolved from running the ship to pushing paper. Hearn spent countless hours in his stateroom logging records, time charts, and data, managing the milk run back and forth to Puerto Rico. It was load and roll. Everyone was hustling. There wasn't time for him to inspect every box that went on his ship and quiz every deckhand. Following the arrests, TOTE hired security guards to search crew as they came aboard.

TOTE's firing of the four officers came as a shock to those working on the vessels. Haley wasn't even on duty when the drugs were found. Two respected captains and two chief mates, all elders of the trade, were gone. With them, decades of knowledge and experience had been tossed out like yesterday's garbage. The message was clear: no one's job—on land or at sea—was safe.

TOTE became ruthless, driven to squeeze as much profit as possible out of an operationally expensive industry. Some mariners who worked for TOTE say that the company was making a significant profit at that time. The ships cost several million dollars, the labor, the berthing, the fuel, the endless maintenance, plus the insurance (*El Faro*'s hull and machinery were covered for $24 million)—all these big-ticket necessities cut into their bottom lines.

And lately, cargo prices had plummeted; worldwide, there was too much capacity, an abundance of ships, and not enough customers. Hanjin, one of the world's largest shipping companies, filed for bankruptcy in 2016, leaving seventy-eight of its laden ships wandering the oceans in search of ports that would unload their goods without guaranteed pay.

Piracy also plagued the industry. Notorious waterways like the Malacca Strait (between Malaysia and Sumatra), the South China Sea, the Gulf of Aden (the entrance to the Red Sea), and both coasts of Africa teem with pirates looking for valuable cargo or, even better, officers to ransom. In late October 2017, six crew from a German container ship approaching a Nigerian port were reported kidnapped. In 2013, the World Bank estimated that the annual cost of piracy off the coast of Somalia ran somewhere around $18 billion.

Cyberattack posed another threat. Shipping is heavily dependent on computers for tracking vessels and complex logistics. Maersk, a global shipping giant based in Denmark, found its data held for a $300 bitcoin ransom (about $1 million at the time) by malware in June 2017. The attack cost the company approximately $300 million in lost revenue.

Compounding TOTE's plight: Puerto Rico, the company's

cash cow, was collapsing. Islanders were fleeing the bankrupt territory, reducing demand for goods from the mainland. TOTE's direct competitor in the Puerto Rican trade, Horizon Shipping, went bankrupt in early 2015.

Further, new environmental regulations were about to render TOTE's old steamships obsolete.

In 2011, TOTE engaged a consultant to figure out how to run the company leaner. He immediately replaced long-standing managers and staff. Other positions were deemed unnecessary and eliminated. It appeared to those who worked on the ships that TOTE wanted younger officers, and the cocaine bust seemed like an excuse to clear house.

TOTE may have had an ulterior motive for firing Captain Hearn: the old shipmaster had become a troublemaker. A few years before the drug bust, TOTE replaced *El Faro* with a new class of vessel in Alaska to meet environmental regulations. The old steamship was tied up in a Baltimore slip. Hearn had mastered *El Faro* in the icy Pacific Northwest for seventeen years, then was transferred to helm her sister ship *El Morro* in the warm waters between Jacksonville and Puerto Rico.

In his estimation, *El Morro* was an inferior ship, a rust bucket, seriously neglected while working herself to death in the corrosive Caribbean run. The hot, humid climate relentlessly gnawed away at her steel, leaving rusty tears streaming down her ancient hull.

Both *El Faro* and *El Morro* were nearly forty years old but to Hearn, *El Faro* was an old friend, superior to his current clunker. It ate at Hearn knowing that she'd been left to rot away in a Baltimore slip.

Hearn's fondness for his ship wasn't rare in the maritime world. Many people who work closely with boats think of them as living beings, complete with their own personalities and peccadilloes. Even something as unwieldy as a cargo ship can have ardent admirers.

In *The Log from the Sea of Cortez*, John Steinbeck wrote eloquently in 1951 of the intense bond we instinctively form with boats: "How deep this thing must be. The giver and receiver again; the boat designed through millenniums of trial and error by the human consciousness, the boat which has no counterpart in nature unless it be a dry leaf fallen by accident in a stream. And Man receiving back from Boat a warping of his psyche so that the sight of a boat riding in the water clenches a fist of emotion in his chest."

Hearn felt that clench in his heart each time he thought about *El Faro*. He alerted TOTE to the fact that *El Morro* desperately needed major maintenance. He considered the ship unsafe. When he wasn't satisfied with TOTE's unresponsiveness, he informed the coast guard.

Shortly thereafter, because of the drug incident, he was out of a job.

Retribution against seamen for lodging safety complaints violated maritime law but that didn't mean it never happened. And sometimes mariners fought back. A few years before Hearn's firing, Captain John Loftus, a shipmaster with forty years of experience, was fired for calling out safety concerns aboard a vessel owned by TOTE's competitor, Horizon Shipping. In 2014, Loftus won a million-dollar whistleblower case against the company. The federal judge assigned to the case agreed that Loftus's concerns were warranted.

Shortly after TOTE fired Hearn, the coast guard inspected *El Morro* and concurred with the captain's assessment: Much of the ship's deck steel was wasted, which made her structurally unsound. Instead of paying for the expensive repairs required by the coast guard, TOTE scrapped her.

While Hearn was lobbying against *El Morro*, TOTE was building a case against the other officers to hold them accountable for the aging ship's deterioration. The three seamen had spent decades overseeing the company's aging vessels' eternal battle against the sea: scraping, painting, and epoxying, washing down decks, cleaning out pumps, compiling repair lists, and hunting down parts no longer in production. Instead of a bonus, they received warning letters from TOTE claiming that they weren't doing enough to maintain the ships. By the time the drugs were found on *El Morro*, TOTE had a paper trail pinning the ship's condition on its senior officers. Termination letters followed.

Subsequent arbitration led to undisclosed settlements in favor of the mariners, but that was scant comfort for those who had given so much of their lives to the sea.

TOTE put *El Faro* back into service on the Puerto Rican run, tag-teaming with her sister ship *El Yunque*. It was a temporary fix. TOTE had just ordered two new liquid natural gas–powered (LNG) ships to replace its two remaining elderly steamships. It would take several years before those new vessels were operational, so the old steamships continued chugging back and forth from Jacksonville to Puerto Rico, patched, painted, and duct-taped together.

El Faro and *El Yunque* were not only aging, they were taking on more cargo since the bankruptcy of competing shipping line Horizon, especially reefers—refrigerated containers that required constant tending to prevent the food inside from spoiling. It was pricey cargo, so it made money, but it strained the already thin crew.

As the ship's air-conditioning struggled against the day's stultifying heat, Third Mate Jeremie let loose his frustration with the increasing workload. One of the reefers hadn't gotten plugged in during loading the night before, he told Jack. Now a whole trailer full of food was spoiled.

A port mate once helped with loading in Jacksonville, he said, but that position hadn't been filled since early September.

Jeremie was sick of the constant scrambling and lack of support. "You know what's changed?" he said to Jack. "I could not fucking keep up with the loading. I had a goodie helping me"—a GUDE, general utility, deck, and engine unlicensed seaman trained in all areas of the ship—"He couldn't keep up. I was helping him plug in and I didn't have time to get all the temps down."

All the reefers got loaded on just before the ship left the dock, creating a shitstorm of problems. They had to scramble to get everything plugged in and secured to leave on time.

"We used to have a port mate and now we don't. We had a longshoreman, now we don't. Then we also lost our electrician. We used to have that system of checks where a guy would come down and make sure that every reefer was good. That doesn't happen anymore."

It was true. Lately, positions on the ships and onshore were being cut, and the remaining crew had to pick up the slack. Setting high standards was tough when the cast of characters was constantly changing.

"Your average union electrician wouldn't even come on a ship like this," Jack said. "It's too much work for them. They're not gonna work their way through jungles of lashing chains and dirt to get to a plug somewhere."

"It's insane down there in the holds," agreed Jeremie. He guessed that the ship was carrying more than three dozen reefers, each one demanding someone's time and attention during the short docking period between voyages. Second Mate Charlie Baird would lay out all the electrical cords neatly when the empty ship was heading back to Jacksonville, but his relief, Second Mate Danielle, didn't do that.

"There's just extension cords everywhere. It's a mess down there. Everything is falling apart. I'm doing what I've always done, but it's just not enough anymore."

Jack nodded his head.

"I don't think I could ever be captain here," Jeremie said. "I'd lose my shit."

Actually, Jack said, it's good to be captain. "The captain and chief mate go down to their rooms and play video games," he said.

"Right," Jeremie said. "These guys got it all figured out."

El Faro was poorly run, they agreed, but the unlicensed crew was okay, Jack said. He would know since he was unlicensed, too.

On the ships, officers and crew rarely talked to each other. The gulf between the two classes of mariners could be as stark as black and white. They had separate unions, separate sleeping quarters, separate mess halls, and separate lives. The officers on *El Faro* lived in New England or southern Florida. As members of the American Maritime Officers union, they could call their hall to get work instead of showing up in person and they usually signed long-term employment contracts.

The unlicensed crew on *El Faro* mostly came from tough Jacksonville neighborhoods, where a seafaring job could save kids from a life in and out of prison if they could resist becoming drug mules. Guys like LaShawn Rivera, thirty-two, who was working as a cook in the galley on *El Faro* on her final voyage.

Shipping had been the young man's salvation. He'd been raised in Atlanta by his mother and stepfather, Robert Green, a bank manager who later joined the ministry. When LaShawn was a teen, the family uprooted to Jacksonville. Away from his cousins and hometown, the boy got teased at school for being different, and he withdrew to the streets. Soon he started dealing drugs and spent time in juvie. Then he had a baby girl.

For a young troubled kid in Jacksonville, few job prospects lay ahead, but he was determined to take care of his child.

One day, an elder of the neighborhood told him about the merchant marine. It sounded like the way out—one of the few good jobs available to folks without a college degree. LaShawn drove over to the Seafarers International Union hall and studied the job board. All those ships heading to exotic places around the world called to him. He spent all day every day at the hall until he landed a spot on a ship.

The first time he sailed abroad, he called his stepdad from London. "I'm never coming back," he told him. "Back," Pastor Green says, meant the Jacksonville streets.

In search of a specialty, LaShawn took cooking courses through his union and eventually was certified to prepare food for the thirty-three men and women on *El Faro*, his dreadlocks loosely tied back. Working on the ships gave him a sense of pride and purpose, but it would never make him rich. He had a fiancée and then another little girl to support on his $80,000-a-year salary. In his spare time, he consumed popular business and self-help books, trying to find a better way to care for his family. When he shipped out on *El Faro* in late September, Shawn left behind his older daughter, plus a one-year-old baby girl, his fiancée—pregnant with their second daughter—and a stack of books promising him a better future.

After sailing for forty years, Jack Jackson had met many guys like LaShawn Rivera and plenty of other unlicensed seamen who were much rougher around the edges. Trapped on a ship way out at sea, things could get ugly, sometimes violent, and you couldn't just call the cops.

"You wouldn't believe the fucking shit that goes on on these ships, ya know?" he said to Jeremie. "I've always had the bad luck of being around these crazy eccentrics. I mean, I've been with some real sickos, man. I mean, some real fucked-up people, man."

Jack's assessment was that during this tour, at least, they were sailing with some good people.

Jeremie looked out at the sweltering sea. Dazzling sunlight burned up the hazy dawn. He was happy to hear he wasn't a sicko.

At 9:30 a.m., Davidson walked in. "What's this silly thing telling us?" he asked, gesturing to the NHC weather alerts coming in through the SAT-C printer.

Jeremie joined him at the machine and studied the new data. "Showing the wind warning and all that stuff. We're going into it."

"We're going into the storm," Davidson said turning to face down the horizon. "And I wouldn't have it any other way."

For Davidson, handling a massive cargo ship was a true test of a man's mettle. Courage was a job requirement for taking the top post in the American merchant marine. You had to be ready to fight drunken seamen, confront recalcitrant officers, and wrangle more speed out of your chief engineer. And there was always the threat of the sea itself. That said, Davidson hadn't yet seen much weather on *El Faro*. Erika was the first real storm he'd been through on that ship. And thanks to Charlie Baird, he hadn't witnessed how *El Faro* handled in a full-blown tropical cyclone. Their route during Erika through the Old Bahama Channel took them south of the islands. They had forty-knot winds on their port side, but were in the lee so the seas stayed calm enough that the crew had been able to work as they steamed through.

"Ship's solid," he told Jeremie authoritatively. "There was no shortage of swell running during that storm."

"This ship is solid," Jeremie agreed. "The hull itself is fine. The plant no problem. But it's just all the associated bits and pieces, all the shit that shakes and breaks loose that's the problem. And where the water goes," he added prophetically.

"Just gotta keep the speed up," said Davidson, perturbed that his third mate didn't simply agree with him the way a third mate should. "And who knows? Maybe this low will just stall a little bit, just enough for us to duck underneath."

Throughout the voyage, Davidson never referred to Joaquin as a hurricane. He called it a "low," or a "storm," or a "system," continually downplaying Joaquin's power, either to reassure himself or to firm his officers' resolve.

Davidson then reminisced aloud about storms past. It was the mariner's way of showing off experience. And balls. He'd earned his stripes: "The worst storm I was ever in was when we were cross-ing the Atlantic. Horrendous seas. *Horrendous.* We went up, then down, then down further. You'd roll a little bit into the trough, then you'd rise up a crest and come crashing down and all your cargo would break loose. The wheels on the vehicles aboard moved so much that the lashings loosened up. We were in that shit for two days. The engineers would be cleaning the fuel strainers every three hours."

On a steamship, keeping the fuel lines clear is critical—they're the main arteries to the boiler. And they can get blocked. Heavy fuel oil, the kind used for older cargo ships, is thick, viscous, and dirty, rife with impurities and solids. Just like a bottle of wine that's been stored for a while, sediment can settle at the bottom of the holding tanks and accumulate there. Sudden movement will stir up that gunk, which then clogs the lines that feed the boil-ers. To minimize this problem, the fuel lines are equipped with strainers—wire mesh trays that look like fryer baskets—and under normal operation, these have to be cleaned out daily.

In rough weather, with the ship tossing and turning, engineers might find themselves cleaning those strainers with greater fre-quency. One former officer told me that several years ago, when *El Faro* was in thirty-foot seas, an engineer had to clean out the strainers every three minutes. It's arduous work, not as simple as dumping fryer oil. The whole fuel strainer assembly needs to be taken apart to access these strainers, then put back together.

Davidson recalled that during the Atlantic storm, the ship's engineers had to slow the turbines to protect their machinery. "We could only do six knots," he told Jeremie. "The winds were 102 knots, force 12." (Force 12 is the maximum level on the tradi-tional Beaufort scale, an empirical measure of wind speed as it relates to sea conditions. The scale was widely adopted in the nine-teenth century as a way to standardize mariners' observations at sea. Force 12 refers to winds greater than 73 miles per hour and greater than forty-six-foot waves. On land, force 12 would cause "devastation," according to the scale. At sea, the air would be thick with sea spray and foam, and visibility would be nil.)

"It was like that for a solid twenty-four to thirty-six hours," Davidson continued. "I shit you not. This for a couple of days. It was bad. It was bad. We had a gust of wind registered at 102 knots. It was the roughest storm I had ever been in."

Jeremie wasn't the kind of man to one-up Davidson with his own sea stories. Ships are supposed to go around storms, not through them, though sometimes they get caught in something and have to fight their way out. But thinking about Davidson's story made him consider *El Faro*, a ship he knew well. She didn't handle storms gracefully. He'd seen it with his own eyes.

"The scariest thing I ever saw was that on this ship you're a lot closer to the water, not far above the waterline," he said. "At night we'd get into a trough and see the white line of the waves breaking right next to us, all the way up here on the bridge."

Davidson was too captivated by his own tale to hear Jeremie. Just thinking back on that Atlantic storm got his adrenaline going. That was real seafaring there. A wild ride. He continued spinning his yarn: "We had a rogue wave on every seventh or eighth wave in a period. On the bridge, all hell broke loose. Before we went through that thing, I would've said no way could the knobs of a radio to get blown off. Well, it happened."

Hell, maybe they'd get those kinds of seas this time. But maybe not. "I think we'll duck down south enough and we'll speed up," Davidson said, just a touch disappointed.

At 10:30, another NHC weather forecast came through NAV-TEX on the bridge. Jack tore the sheet off the printer. "It's moving away fast," he said, scanning the coordinates.

"No," said Jeremie looking at it more closely. "It's not moving away, not yet."

He walked over to the weather chart they'd been working on

and plotted out the latest forecast. They were heading southeast, skirting the islands on the Atlantic side, and Joaquin was heading southwest, right for them.

"We're on a collision course with it."

At the time, the winds were only blowing 55 knots at the hurricane's center.

When Davidson left the bridge, Jack's mind wandered to the seafarers of long ago in their sailing ships, crossing the Atlantic, encountering a hurricane for the first time. "I wonder what the first Spanish sailors, those that survived, thought when the eye passed right over them," he said. "They're thinking, *It's over.*"

The two men laughed. When the center of the hurricane passes above you, it's eerily calm; often, you can see straight up to blue sky. It's easy to think the hurricane has miraculously cleared up. In truth, the worst is yet to come as the eye wall bears down.

"I think back then," Jeremie said, "they didn't know the difference between a storm and a hurricane. They figured a hurricane was just a really bad storm. All those ships that sunk, the Spanish probably said, 'sunk in bad weather.' *Probably a goddamn hurricane.*"

"No one believed them," Jack said. "The survivors would talk about hurricanes back in Spain, and people would say, *Yeah, yeah. Sure, sure. We know what storms are.* No you don't know what this storm's like."

It's true. The first Spanish explorers in the New World had no frame of reference for such deadly tempests; there aren't cyclonic storm systems like that in the Mediterranean or along the Spanish coast. The earliest European descriptions of hurricanes emerged a decade after Christopher Columbus's initial voyage across the Atlantic from the Canary Islands to the Bahamas.

Since arriving in the region in 1492, Columbus had learned about the spiraling storms that came every July through October from the Taíno, the native people of Puerto Rico and surrounding islands. The Taíno called these storms *juracánes*—acts of a furious eponymous goddess, which they depicted as a disembodied head adorned with two propeller-like wings. She looked remarkably like the hurricane symbol you find on modern maps.

Columbus was sophisticated enough, or frightened enough, to use tools the Taíno had given him to predict the approach of a *juracán* as it advanced toward Santo Domingo in the Dominican Republic. He observed cloud formations and the swell of the sea and warned all to secure the city's main port. His small fleet sheltered in place in a bay while the Spanish governor, dismissing the warning as witchcraft, sent some twenty-six ships laden with gold to their peril.

In their attempt to reconcile the hurricane with biblical or classical references, Spanish chroniclers of the sixteenth century came up empty and called it the devil's work. But actual mariners had a vested interest in understanding these unprecedented tempests, in spite of the church's tenacious hold on apparent truth. Juan Escalante de Mendoza, captain general of the New Spain fleet, covered hurricanes and their dangers in his mariner's guide published in 1575, writes Stuart Schwartz in his book, *Sea of Storms: A History of Hurricanes in the Greater Caribbean from Columbus to Katrina.*

Mendoza called them "a fury of loose contrary wind, like a whirlwind, conceived and gathered between islands and nearby lands and created by great extremes of heat and humidity."

This account reveals a surprisingly accurate understanding of the storm's mechanism. Mendoza also noted signs of an imminent hurricane, including odd behaviors among birds, which are exquisitely sensitive to shifts in barometric pressure. When they sense a pressure drop, they might try to outrun it. The result is unusual species showing up in unexpected places, sometimes migrating from America to Europe to escape a hurricane. A few days before a storm, birds might go into a feeding frenzy to bulk up before winds and rains wipe out their feeding grounds. All these behaviors are easily observable.

Schwartz writes that Mendoza was "careful to mention that 'the things that are to come, you know, sir, only God our Lord knows, and none can know them unless it is revealed by His divine goodness.' Prediction of the weather always treaded dangerously close to the Church's disapproval of divination."

CHAPTER 5

A HURRICANE IS NOT A POINT ON A MAP

A hurricane is not a point on a map.

It is not an object that exists in space and time. Rather, it's a huge catharsis—a brief, explosive event when nature's forces combine to spin off the ocean's heat into wind. Over its brief life span, a hurricane expends the power of ten thousand nuclear bombs. It's a spectacular display of thermodynamics in a complex, evolving, moving system.

Like a cancer, a hurricane is a lethal distortion of the stuff of everyday life. The earth's winds harmlessly swirl about us, creating patterns in the clouds, kicking up waves for surfers, and nudging planes around the planet. All these heavenly movements are tendrils of much larger systems, like the jet streams that forever flow east, caught up in the rotation of Earth. Thanks to the jet stream, winter storms always pound New England in February when arctic air from Canada drifts south over the plains, gets swept east in its current, and is carried to the warmer air over the Atlantic Ocean.

Although the earth's meteorological picture is as vast as the planet itself, it is very sensitive, prone to disruption, fickle. A butterfly-wing-like change in pressure or temperature in one place can cause a small piece of the continuum to break away from the mainstream like a recalcitrant teen. It often fades into the ether. But sometimes, when conditions are right, it can escalate exponentially, causing astounding damage to anything in its path.

Perhaps to domesticate these mighty systems, we give them names—Katrina, Sandy, Mitch, Joaquin. We mark them on our maps and say *this is where she is*. We draw a line and say *this is where he will go*. As if one day, a storm named Katrina rose out of the Gulf of Mexico like a Japanese monster hell-bent on ravaging New Orleans. As if you could put a beacon on her and track her every move as she made her way to her target. As if there were some kind of motive behind the destruction.

```
TROPICAL DEPRESSION ELEVEN FORECAST/ADVISORY
NUMBER 1: 0300 UTC MON SEP 28 2015: TROPICAL
DEPRESSION CENTER LOCATED NEAR 27.5N 68.7W AT
28/0300Z. PRESENT MOVEMENT TOWARD THE NORTH-
WEST AT 2 KT. MAX SUSTAINED WINDS 30 KT WITH
GUSTS TO 40 KT.
```

Every hurricane begins as an atmospheric low, or depression, in the bottom layer of the earth's atmosphere—the troposphere—which reaches up to seven miles above the planet's surface. The low acts like a vacuum, pulling up warm, moist air from the ocean, which spirals around it in a counterclockwise direction in the Northern Hemisphere. If conditions are right—plentiful warm, humid air—the currents corkscrew upward around the low pressure zone at increasing speeds. When that heated air hits the much cooler upper atmosphere, it condenses, shooting out of the hurricane's top like a whale's spout, away from the center, and falls to earth as rain.

Feeding on the temperature and humidity differential between the hot ocean and the chilly upper atmosphere, the tropical storm thrives. Air currents around the center pick up heat, and with it, speed, as the center's pressure drops even lower, intensifying the cycle. "Warmest climes but nurse the cruellest fangs," wrote Herman Melville.

Joaquin was born as a tropical depression off the Canary Islands, three thousand miles east of Puerto Rico, a birthplace strange and rare for tropical cyclones because it was so far north. Designated Tropical Depression Eleven, it remained a loose cluster of showers that meandered across the North Atlantic toward the Caribbean. Forecasters at the National Hurricane Center in Miami ran their computer models and concluded that the system would dissipate. Over the course of a few weeks that September, however, Eleven defied the odds to become a cohesive system.

Even as Joaquin matured, dozens of computer models at the NHC in Miami predicted its demise. Some didn't, but they were dismissed as outliers. On Monday, September 28, the day before *El Faro* departed from Jacksonville, the NHC issued an advisory: "The forecast for T.D. Eleven is to maintain tropical depression status while drifting slowly NW and will likely dissipate by the end of the week due to unfavorable winds aloft." When Joaquin failed to comply, it took NHC forecasters by surprise.

In spite of a moderate shear—competing crosswinds that most forecasts predicted would blow it apart—at midnight on September 29, Eleven evolved into Tropical Storm Joaquin.

TROPICAL STORM JOAQUIN/ADVISORY NUMBER 6:
0900 UTC TUE SEP 29 2015: TROPICAL STORM CEN-
TER LOCATED NEAR 26.6N 70.6W AT 29/0900Z.

```
PRESENT MOVEMENT TOWARD THE WEST AT 4 KT. MAX
SUSTAINED WINDS 35 KT WITH GUSTS TO 45 KT.
```

Throughout Joaquin's evolution from depression to hurricane, the NHC could only guess at how it would develop, and what path it would take. The center issued discussions—carefully worded explanations of its forecasts of the storm system's path and intensity. Embedded within these discussions were clear admissions of ambivalence: "There is considerable uncertainty among major models with the details of track . . . intensity and timing not only of Joaquin, but also the surrounding environment," the center wrote at 2:47 p.m. EDT on Tuesday, September 29.

The same uncertainty was included in its weather discussion issued thirteen hours later.

Uncertainty in forecasting can be quantified. When models contradict one another, uncertainty is expressed as a probability.

In the course of everyday life, we don't often encounter uncertainty. Gamblers and hedge funders may weigh odds all day, but most of us aren't sure what to do if someone says she's 30 percent sure she's wrong. How do you process that, especially in a world where so many decisions are made for us by technology?

"If the definition of wisdom is understanding the depths of your own ignorance, meteorologists are wise," says Kerry Emanuel, an MIT professor who has dedicated his life to understanding weather and climate. "It's wise but it's a wisdom that is not recognized. If you say there's a lot of uncertainty in this, in the modern world, it's translated as *You don't know anything.*"

Due to uncertainty, prudent mariners follow the 3-2-1 rule: Three days ahead of a hurricane's forecasted position, stay three hundred miles away; two days ahead, keep out of a two-hundred-mile radius of its projected center; one day ahead, stay one hundred miles away from its eye in all directions. The rule is based on the fact that hurricane paths are erratic and unpredictable, so it's smart to give the system a wide berth.

But mariners often need to make binary decisions based on nebulous weather forecasts. On October 24, 1998, the elegant *Fantome*, a 679-ton staysail schooner built in 1927, departed Honduras for a six-day windjammer cruise. A thousand miles away, Hurricane Mitch rumbled in the Caribbean Sea. As Mitch picked up strength, the captain of the *Fantome* got nervous and discharged his passengers in Belize City, then headed north toward the Gulf of Mexico to outrun the storm.

Forecasting Mitch proved extremely difficult due to weak steering winds, but the official NHC prediction, issued with multiple caveats, was that the storm would go north toward Mexico's Yucatan Peninsula. When the *Fantome*'s captain received that forecast, he hove to and headed south, unwittingly right into the hurricane's path, which, contrary to forecasts, took a left turn toward Central America. On October 27, fighting 100-mile-per-hour winds and forty-foot seas, the *Fantome* was lost forty miles south of the hurricane's deadly eye wall.

Slow and unyielding, the Category 5 storm's winds and rains killed more than eleven thousand people in Honduras, Nicaragua, El Salvador, and Guatemala, making it the second-deadliest storm in the Atlantic's history.

```
HURRICANE  JOAQUIN  FORECAST/ADVISORY  NUMBER
11: 1500 UTC WED SEP 30 2015: HURRICANE CENTER
LOCATED NEAR 24.7N 72.6W AT 30/1500Z. PRESENT
MOVEMENT TOWARD THE SOUTHWEST AT 5 KT. MAX
SUSTAINED WINDS 70 KT WITH GUSTS TO 85 KT.
```

Weak, meandering, dispersed Joaquin was precisely the kind of storm the NHC has trouble forecasting, says James Franklin, director of the center, as we sit in his Miami office a year and a half later. James has two MIT degrees, rimless glasses, and a quiet, analytical manner. He speaks in complete paragraphs. But his quiet demeanor belies an intrepid soul. James used to fly in NOAA Hurricane Hunters, straight into tropical storms. The Gulfstream IV is a high-altitude jet that flies in and around hurricanes recording conditions and dropping sondes (disposable devices outfitted with a parachute) that gather critical data about the storms as they fall from the sky. The planes often ride hurricane updrafts and then, on the very inside edge of the eye wall where the air all rushes down, plummet more than nine hundred feet.

"Have you ever been on the Tower of Terror ride at Disney?" he asks me. "It's basically a big elevator where you get dropped. So it's sort of like that. I did that for seventeen years."

Joaquin just didn't look like it meant business, until it did. "By getting the intensity forecast wrong," James says, "that contributed to our getting the track forecast wrong. If we had correctly anticipated that Joaquin was going to beat the shear and remain a stronger storm, that would have argued for a forecast more to the south."

James explains why even with advanced computers and significant amounts of data, we still can't accurately predict the future 100 percent of the time. In fact, an important part of the NHC meteorologists' job is to keep tabs of, and learn from, error. The NHC is one of the few government offices that obsessively tracks its own mistakes; meteorologists use these errors to improve their models and methods.

On the NHC website, there's an entire section dedicated to its own errors going back to 1970. Multiple line graphs detail two kinds of errors: track errors and intensity errors. What path the storm takes depends on the larger forces around it. Like a leaf floating down a powerful river, it can get caught in swirling eddies, loop back, cut loose, and stall in slower currents near the riverbank. Meteorologists often think of weather systems in terms of fluid dynamics. But unlike a leaf, the hurricane is a nine-mile-high engine—making its movements and behavior even more difficult to predict.

Heat is the fuel that drives the hurricane's engine. The more heat the system absorbs into the upper atmosphere from the warm waters below, the faster it spins, converting the heat's energy into powerful winds. That's why hurricanes in the Northern Hemisphere occur late in the season, after the summer sun has warmed the southwestern Atlantic up to 84 degrees.

As oceans continue to warm due to climate change, hurricanes will get more intense because they've got more fuel to convert to energy. We've already witnessed proof of this; several tropical cyclones have broken records in the past decade alone. Hurricane Sandy, which pounded the New Jersey and New York coasts in October 2012, was the largest Atlantic hurricane on record, spanning eleven hundred miles. A year later, Typhoon Haiyan in the Philippines became the strongest tropical cyclone to hit land ever recorded, with one-minute sustained winds recorded at an astounding 195 miles per hour, killing at least sixty-three hundred people. Hurricane Patricia intensified at an unprecedented rate off the western coast of Mexico two years later, churning out record-breaking maximum sustained winds of 215 miles per hour.

Hurricane Irma, followed by Hurricane Maria, revealed exactly how destructive these systems can be. The huge Category 5 hurricane blasted through Puerto Rico on September 20, 2017. Its tornado-like winds knocked out the entire power grid and nearly all cell-phone infrastructure and cut a swath of devastation, peeling off roofs, stripping away all vegetation, and causing widespread flooding. Ports were closed, and much of Puerto Rico's shipping infrastructure was compromised by winds and flooding.

In the aftermath, Puerto Ricans found themselves without fresh food, water, and electricity. Hospitals, schools, homes, and factories went dark. Some towns were completely wiped out. Tankers were unable to dock or unload; without fuel, emergency efforts ground to a halt. By mid-October, more than 1.2 million people on the island lacked access to potable water.

```
HURRICANE  JOAQUIN  FORECAST/ADVISORY  NUMBER
13: 0300 UTC THU OCT 01 2015: HURRICANE CENTER
LOCATED NEAR 23.8N 73.1W AT 01/0300Z. PRESENT
MOVEMENT TOWARD THE SOUTHWEST AT 5 KT. MAX
SUSTAINED WINDS 100 KT WITH GUSTS TO 120 KT.
```

Joaquin would continue this superheated superstorm trend with its own series of unprecedented stats. As it was forming, it traveled westward, but a ridge of high pressure blocked its forward progression, redirecting it south over the hot Bahamian waters—about 86 degrees—two degrees hotter than ever recorded at that time of year. This was the heat Joaquin needed to thrive. For sixty hours, the storm fed off these warm waters and rapidly intensified. At 2 a.m. on September 30, about 170 miles east-northeast of the small Bahamian island of San Salvador, Joaquin became a hurricane.

Mariners live and die by weather. Naturally, they are obsessed with forecasts, signs, and augurs. Long after a seaman has turned his back on the ocean, he follows the weather. He might live in landlocked New Mexico, four hundred miles from the nearest boat, but he'll still be able to tell you all about the system forming off the coast of the Bahamas.

At home in Portland, Maine, on September 29, 2015, *El Faro*'s off-duty Second Mate Charlie Baird couldn't pull himself away from the Weather Channel. He'd just stepped off *El Faro* a week before, relieved by Second Mate Danielle Randolph. The pair had been tag-teaming the job for a couple of years, seventy days on, seventy days off. Now he was on his sofa, still in his robe at ten o'clock in the morning. He didn't like what he was seeing, this storm developing in the Atlantic. He was off the clock, but he couldn't help himself.

He picked up his phone and texted *El Faro*'s captain, Michael Davidson: "Storm forming north of the bahamas!!"

Growing up in Portland, Charlie loved the water, loved sailing, loved the sea. In high school, he was a state-champion swimmer, breaking record after record, and got recruited to swim at Niagara College. At sixty-six, he's still remarkably fit, and still competing in triathlons. A handful of medals hang from the chandelier in his dining room.

Charlie joined the merchant marine in the early '80s. He's been shipping so long that, like many mariners, he's missed a lot of life while at sea. Like someone who's spent a few years in a coma, he's kind of disconnected. A lot of casual conversation doesn't make sense to him—movies, music, and events have simply passed him by. He would make a terrible Trivial Pursuit partner.

Charlie lets this all wash over him as he swigs another glass of red wine on ice.

He knew not to mess with hurricanes. First, there are the winds. At low speeds, the wind's force on a large ship is minimal. That is, up to about 35 miles per hour. Wind speed and force have an exponential relationship, meaning that as the wind notches up, its force doubles, then triples, and then quadruples, and so on. It's based on a simple formula: wind pressure per square foot = 0.00256 x (wind speed)^2. Put another way, between 1 mile per hour and 35 miles per hour, the wind's force against the boat runs roughly six to nine times its velocity. That's the kind of incremental change that the human brain can handle. But over the next 35 miles per hour, its force rapidly grows to eighteen times its velocity. At 112 miles per hour, the wind's force is thirty times its velocity—you can't stand upright in that kind of wind no matter how far you lean; first you'll slide across the ground, then you'll get thrown backward into an uncontrollable somersault.

This power makes a big difference when you've got a crosswind across a large surface area, say, like the side of a container ship piled with boxes. At midspeeds, the wind's force would be equivalent to twenty-one tons pushing up against that ship. You'll heel a bit, and might want to shift your ballast to try to correct. Ratchet up the wind's velocity to 110 miles per hour, and you've got about 375 tons trying to push the ship over. That's the cumulative force of three of the world's largest locomotives. If you've got vulnerabilities, like a low deck with lots of openings, that kind of force can be catastrophic.

And then there are the waves. At first, the cyclone's high winds cause whitecaps to form on the water's surface. As the hours pass, the waves grow. Start with a completely calm sea at noon on Monday; a 28-mile-per-hour wind starts to blow and by 11 p.m. on Tuesday, you'll have thirteen-foot waves. A rough guide for wind to wave height is two to one—60-mile-per-hour winds (52 knots) stir up thirty-plus-foot waves, depending on fetch, or the length of open water over which the wind blows.

Of course, most mariners hope they're never caught in seas like that, especially in a small boat.

How big can ocean waves get? For millennia, sailors spun stories about extreme waves a hundred feet high that came out of nowhere and smashed apart boats with just one hit. These were dismissed as salty tales, like mermaids and sea monsters. Even in the modern era, scientists considered freak waves a physical impossibility, in spite of compelling evidence to the contrary. During World War II, a rogue wave nearly sunk the *Queen Mary* ocean liner. The thousand-foot-long luxury passenger ship had been stripped down to carry American troops to Europe, and in December 1942, it was loaded with more than sixteen thousand American soldiers when a ninety-foot-high wave smacked her broadside, causing the ship to roll 52 degrees. There were plenty of witnesses to vouch for that one.

It wasn't until modern equipment began recording rogue waves that scientists finally acknowledged their existence. On January 1, 1995, a laser aboard an oil rig in the North Sea recorded an eighty-four-foot-high wave. Since then sea buoys, satellites, mariners, and passengers have recorded these waves, confirming that indeed, they're no myth.

Twenty-two minutes after Charlie texted him about Joaquin, Davidson texted back: "yup . . . thx for the heads up."

Charlie sat on that sofa all day, watching the weather.

Monitoring the storm was the second mate's job.

Danielle should've been obsessing about it just like him.

Charlie knew she wasn't. He was fond of his fellow Mainer, a perky, freckled redhead half his age. He'd taken her under his wing like a little sister a decade ago when she joined the company while a cadet at Maine Maritime Academy. Back then, she was all energy, but ten years of shipping was wearing her down. Danielle recently found out she'd been passed over for a promotion to the new LNG ships. That killed what was left of her enthusiasm for her shipping career, and now he thought she was clocking in, clocking out.

Basic tasks seemed to elude her, like correctly plotting the ship's location during her watch. She didn't actively seek extra work to keep things running smoothly on the decrepit ship. Why bother? TOTE had kicked her to the curb after she'd sacrificed so much of her life to this career. Danielle started taking over-the-counter meds to fall asleep during her breaks and slamming caffeine to stay awake during her watch. These days, she went unusually quiet when her closest friends in Maine asked her about her life at sea.

And there was more. As Charlie was preparing to leave the ship to go home this last time, Danielle pulled him into her cabin and told him something about Captain Davidson he didn't want to hear.

There was no way she was thinking about weather.

But Charlie was.

He couldn't take it anymore. At 6:31 p.m., he texted Davidson again: "Whats your plan?"

"we'll steam our normal direct route to SJP," Davidson texted back a few minutes before 7:00. "no real weather to speak of until the evening of the 30th. all forecasted information indicates Joaquin will remain north of us and by the morning of the 01st we will be on the backside of her. we schedule to depart the dock at 20:00 tonight so everything is shaping up in our favor."

"Cool if u have to we have routes thru mauagiez crooked isle or ne prov chnl," Charlie reminded him. This was shorthand for the channels between the islands along the way deep enough that *El Faro* could use them as escape routes to the lee side if anything should brew up out there.

"Watchin sox presently up 1-0 over ny," he wrote.

"2-0 sox," Charlie added a few minutes later.

"go sox . . ." Davidson replied.

CHAPTER 6

SECOND MATE DANIELLE RANDOLPH

27.37°N -77.43°W

Just before noon, Second Mate Danielle took over Jeremie's watch on the bridge. She'd worked from midnight to 4:00 a.m., gone to bed, showered, eaten, and was back up reporting for duty. Jeremie walked her through the new course Davidson had plotted. He showed her the hurricane's position on the chart and warned her that it was moving southwest at a very lazy, very unusual five knots.

Danielle knew that getting close to any storm was plain stupid. Up on the bridge rolling in high seas, you were bound to get seasick sooner or later, regardless of how experienced a mariner you were. Danielle didn't like her captain's glibness, either, considering the seriousness of their situation. "He's telling everybody down there, *Oh, it's not a bad storm. It's not so bad. It's not even that windy out. I've seen worse,*" she told Jeremie.

He rolled his eyes and kept quiet. What else could he do? He was only third mate. He walked out to get lunch and unwind before his evening shift.

Danielle turned to her helmsman, Jackie Jones, a Jacksonville man in his late thirties with five children below age twenty, and continued her rant: "He's saying, *It's nothing. It's nothing!* And here we are going way off course. If it's nothing, then why the hell are we going on a different track line? I think he's just trying to play it down because he realizes we shouldn't have come this way. He's saving face."

Davidson showed up a few minutes later, disappointed that they weren't moving as fast as he'd hoped. They were steaming at a healthy 18.9 knots. "Damn, we're getting killed with this speed," he said inexplicably.

"I think now it's not a matter of speed," Danielle warned him. "When we get there, we get there, as long as we arrive in one piece," she added.

The prospect of sailing into a hurricane filled her with dread, which she tried her best to laugh off once she and Jackie were alone. Humor was Danielle's way of dealing with things beyond her control. She turned on Sirius XM radio and sang along to contemporary hits, then switched over to Dr. Laura on Fox News.

"Two of the Polish guys were standing there when he said that," she told Jackie, referring to the five extra hands hired to install heating coils on *El Faro*'s ramps and decks so that trailers wouldn't slip and slide on the ice when the vessel returned to the Pacific Northwest. It wasn't unusual to have extra workers fixing things while *El Faro* was at sea. A ship in port is a ship losing money. "I wanted to make sure they knew the word *hurricane*. One of them smiled and said, *Hurricane! Yes!*"

Danielle laughed. "They're all excited about it. Ah, if they only knew. I remember going through a couple of storms on the *El Morro*. Shit was flying everywhere."

At 12:24, Danielle saw a thick black line on the horizon—the bar of the storm. "There's our weather," she said ominously. "The storm gets really big on the third or the fifth of October," she added, studying the forecast. "Warning in red for this weekend for rain—severe flooding and rain. They never mentioned Maine, though. They didn't say, *Two more days and then it'll cover the entire state of Maine.* Yeah, you know, some people live up there."

Like Davidson, Danielle had grown up on the coast of Maine. Her parents were retired navy; her mother worked as a hairdresser, her dad as a handyman. Danielle's grandmother and great-aunt had emigrated from France after World War II. The older women used to converse rapidly and loudly in French. Danielle understood what they were saying, but she never got the hang of speaking the language.

Danielle spent her childhood in Rockland, a small town on Penobscot Bay, but life wasn't easy for her as a girl. She wanted to belong, wanted to feel a part of something, but her world felt fragmented. Her family's foreignness added to Danielle's feelings of isolation. Danielle lived in a tiny second-floor apartment with her grandmother, separated from the rest of her family—mom, dad, and brother—who lived in an apartment below. Danielle's best friend says it was strange, this little girl and older woman sharing a closet-size room, just big enough for a bed and a radio.

During summer breaks, Danielle worked on the docks of Port Clyde, and at the O'Hara Corporation—a fishing consortium and marina in Rockland. That's where she discovered her love of boats and the ocean. She'd jump in a skiff and paddle around while her great-aunt watched from the shore.

Danielle was drawn to the sea at a very early age and in high school, informed her mom that she planned to go to Maine Maritime Academy (MMA), one of five American maritime academies established by the federal government to train ship's officers and engineers for the merchant marine. Getting into Maine Maritime focused Danielle, gave her purpose.

Maine Maritime was based in the tiny town of Castine, twenty miles northeast of Rockland as the crow flies. If it weren't for the islands in Penobscot Bay, Danielle would be able to see straight across to the school from her house.

Castine once held a strategic spot at the mouth of the Penobscot River—the main thoroughfare for the fur and timber trade in the seventeenth and eighteenth centuries. The town was first settled by the French in 1613 as a trading post. In less than a year, the British seized it, then the Dutch came in, then the British reclaimed it, and finally, the Americans incorporated the town in 1796, though it was contested ground again during the War of 1812.

Now the only people fighting over it are real estate agents. Come summer, wealthy folks "from away" arrive to air out their cottages and put their sailboats in the water. The two kinds of people—full-time Mainers and those "from away"—rub shoulders at T & C Grocery, Dennett's Wharf Restaurant, and Eaton's Boatyard. Mainers keep to themselves. Summer people pay taxes on their pricey properties. It's a practical arrangement that works all the way up the coast.

In winter, when frigid winds whip across Penobscot Bay up the steep streets of the tiny town, Castine's white clapboard houses huddle together against the cold; along the waterfront, small nineteenth-century brick warehouses hint at the town's historic commercial past. This was the only place Danielle wanted to be. Students at MMA could study deck operations and serve on a ship's bridge, or study engineering and serve in the engine room. Danielle wanted to join the ranks of the former. She applied to MMA as a senior in high school and refused to consider any backup schools. For Danielle, it was Maine Maritime or bust.

Danielle's first year at academy was stressful; her hair fell out in clumps. Laurie Bobillot, her mother, remembers seeing bald patches in her daughter's wavy locks. Still, five-foot-nothing Danielle wouldn't quit.

Being a small woman in a man's field had inherent dangers, but her mostly male academy classmates were supportive. As a cadet, however, Danielle had to confront the reality of shipping out with men. During one of her first tours on one of TOTE's vessels, the chief mate called her into his stateroom. He said that he wanted to show her something. She knocked on his door; he opened it and dropped his towel. She fled, told her friends, and word got up to the captain. The chief mate was unpopular with the other officers and was quickly fired for sexual harassment. When the second and third mates moved up to take his place, Danielle was given the third mate position aboard that vessel.

After that, she was branded as *the girl who got her job because she claimed some chief mate harassed her.* Which wasn't quite fair. At the time, she was very young and very green. She did what she was told. She didn't ask for this, she didn't want it. She'd worked hard to be a deck officer in the merchant marine. She wanted to be taken seriously.

Academically, things didn't always come easily to Danielle. She sometimes transposed numbers and dreaded taking azimuths at sea. At two o'clock on Wednesday afternoon as they headed to Puerto Rico during the final voyage, she sent a weather report to the National Weather Service—a voluntary spot-report requested of all commercial vessels at sea—and ended up sending erroneous coordinates that positioned *El Faro* squarely atop mainland Cuba. Because of that mistake, her data was chucked.

One former ship's officer told me that he'd tell her to do something and then she'd walk away and forget it entirely. "Write stuff down," he admonished her. He liked her a lot, everyone liked her, but she was often distracted. It took her a few times to pass her second mate's exam.

Danielle loved life, though, and onshore, she was a regular Martha Stewart—cooking, decorating, planning parties. She had a passion for the past, especially vintage clothes. She would order '50s dresses off the web whenever the ship had internet service and came bouncing down the gangway to dozens of boxes of things she'd bought for herself and friends; it was like Christmas every seventy days.

She adored holidays and planning parties, too. One fall, she decorated a backyard path that led out to a table under a tree with homemade lanterns—mason jar lanterns with votive candles, which she also hung from the tree's branches—and she decked out each setting with decorative cloth placemats and a carved pumpkin with a candle in it. She went all out.

When she was ashore, Danielle was tireless. "She exhausted me," her friend says with a smile. "She had lots of energy, so by the time the ten weeks were over on this end, we needed a break. She would run us ragged because she'd have so much that she wanted to squeeze in."

Danielle's brutal work schedule on the ship, along with a steady diet of cafeteria food and lack of sleep, was beginning to show. She worried a lot about her weight; it didn't help that the other officers sometimes teased her for putting on the pounds. During shore leave, she became a workout fanatic, and built a strong friendship with her fitness instructor, Korinn Scattoloni. Danielle was memorable because she pushed herself hard, enough that Korinn sometimes worried about her. "I watched that red face to make sure it didn't turn purple," she says. At first, Korinn didn't understand why the high-energy woman in the back of the class with the hot pink workout clothes would vanish for months at a time. Eventually, Danielle told her instructor that she was in the merchant marine, a career path Korinn understood well.

There was a time when Korinn herself had answered the call of the sea. She was a successful dancer in New York City when, one day, she went down to see the tall ships at South Street Seaport. As soon as she saw the elegant historic sailing vessels moored to the dock, she knew that was where she was meant to be. Korinn quit her job and signed up to work on the HMS *Rose*, a replica of a twenty-gun Royal Navy ship built in 1757. As a hired hand aboard the three-masted frigate, she was expected to haul and furl the sails, keep the decks and sails clean and mended, and help in the galley.

Late one afternoon, she and another member of the crew were ordered to climb seventy feet up the mast to furl a sail. They'd been working all day and were exhausted, but as the only woman aboard, she wouldn't complain. At the time, they didn't tie-in, trying to be as authentic as possible. Leaning over the yardarm, Korinn began pulling up the yards of canvas, her feet supported only by a slack horizontal rope. When her partner accidentally kicked the rope out from under her, she fell. She would have died if it hadn't been for the chief mate, Robin Walbridge, who violently shoved her into the water right before she hit the deck. She made out with a few cracked ribs and a broken wrist.

Robin later became known as the captain who took the replica of the HMS *Bounty* out to sea during Hurricane Sandy and died with the ship when she was swamped by the storm.

Shipping out all the time made it difficult for Danielle to keep up relationships, which she valued more than anything. She was gone half the year and began to realize how much she was missing. "She was at a point in her life where she was ready to move on," her friend says. Danielle wanted to come ashore for good; the ten-week rotation was dragging her down.

"Yeah, she was missing things," her friend says. "She wanted to be home. She struggled with scheduling, you know? We'd get the calendar out and say, *What Christmas are you gonna be here? What Easter are you gonna be here? What summer are you gonna be here?* That sort of thing, I think that part was tiring."

Danielle was home throughout the summer of 2015. She

threw parties at night, built bonfires, and made s'mores with giant marshmallows. She lazed about in hammocks with her friends and used an app to look at the stars, picking out constellations. Danielle had spent her entire life laughing off things that bothered her. That summer, she laughed a lot.

Still, the specter of another voyage loomed. She refused to talk about life on the ship, but occasionally let her closest friends know how frustrated she was with how much *El Faro* was being neglected. The four men who'd been fired a few years ago, officers who'd trained and supported Danielle, had taken great pride in keeping the old vessel looking her best. Now no one seemed to care.

But like many mariners, Danielle had developed an addiction to shipping. You're addicted to the sea, or you're addicted to the money, or you're addicted to both. You leave your family for months to drive the huge ships on the open ocean. It's intense work and the money's good. You can't find that kind of rush in an office cubicle. You tell them onshore, *I'm quitting in a year. I'm quitting in five years. I'll find a job on land.* But you get into a rhythm, the rhythm of the sea, and you keep going.

Danielle dreamed of coming home for good, but starting a new career would cost a lot of money that she didn't have. Everything she wanted to try required more schooling, which she couldn't afford. The pay aboard the ships was too good to walk away from; she wasn't in a position to start from scratch. She had to keep shipping.

Adding to her frustration that summer was the fact that she was about to lose her home. For years, she'd been living with her two cats in her mother's Rockland house. But now her mother, who'd moved to Wisconsin with her boyfriend, told Danielle that she was selling it. The whole situation upset Danielle. Her grandmother and great-aunt had passed away; her mother was halfway across the country pursuing a new life. Her home, her only anchor, was slipping away.

Late in the summer, Danielle learned that she hadn't been assigned to the new LNG ships. Instead, she was slated to ship out to Alaska with the decrepit *El Faro*. And Captain Davidson.

There was something else bothering her that summer. During her previous tour of duty, in the spring of 2015, it happened again.

Davidson, a man she never liked, came onto her. He caught up with her when she was alone, cornered her, and said, *Will you be my special friend?*

Many guys had hit on her during her career. She was a tough girl, her friends say. She could handle it. But when the ship's captain wants to sleep with you, how do you deal with it? How do you delicately push him away without damaging his ego? How do you escape his advances trapped on a ship at sea? What if you pissed him off so much that he got you fired and ruined your professional reputation? It could happen.

She knew exactly how things played out when a woman reported sexual harassment on a ship. Yeah, she could handle it. But behind her back, the guys would wonder what she'd done to lead on the captain. In cases like these, the woman never escapes unscathed.

CHAPTER 7

COLLISION COURSE

At noon, Danielle and helmsman Jackie Jones watched the sky turn a blinding blue. It was a tease. The hot Caribbean waters simmered in the sunlight but beyond, a storm was lingering, *festering* as Davidson had put it, intensifying off their port bow. It was out there.

The NHC weather report came in around 12:30, sending the printer on the bridge into a temporary frenzy. No change. Joaquin was still lumbering south-southwest, gaining force, with maximum sustained winds now predicted to reach 80 knots by midnight. The storm would be worse than originally thought.

Engineer Jeff Mathias came up to the bridge to check in. He'd been working all morning with the five Polish contractors hired to get the ship ready for the cold Alaskan waters.

Jeff had served as a chief engineer on other ships for several years after graduating from Massachusetts Maritime Academy, but with three young kids at home, he preferred to work on land. In lieu of that, this short, temporary, and lucrative job managing the Polish riding gang on *El Faro* fit the bill.

"Look at you," Jeff said to Danielle, gently teasing his fellow New Englander. "All freshened up, huh?"

She laughed and offered him coffee.

For Jeff Mathias, there was nothing like being at sea and the good income that came with it. But since the birth of his first born, he struggled to balance his passion for shipping with his home life. Overseeing the Polish workers was a good compromise because Jeff could do short stints at sea. He'd worked out his shipping schedule earlier that month so that he could be home for his daughter's first day of school and was scheduled to fly home from San Juan that weekend.

Jeff's ultimate goal was to turn his family's struggling cranberry farm into a destination. In the fall during harvest time, elementary school groups visited, usually guided by Jeff's mom, to learn more about the ancient craft of growing the fruit, first introduced to Massachusetts's English settlers in the 1600s by the Algonquins. Loaded with vitamin C, cranberries played a critical role in early colonial shipping, consumed by New England's whalers and sailors to stave off scurvy during long voyages.

Jeff had spent much of his shore time building an elaborate play castle and maze to attract more visitors to the farm's cranberry bogs. He hoped this would generate additional income to help fight off encroaching real estate development. Just beyond the playhouse and maze, behind a copse of pines, a cul-de-sac of new homes had sprung seemingly overnight. The houses were nearly done, but the dirt between the Mathiases' land and the development was still stripped and raw.

Danielle respected Jeff and his opinions and was relieved to have someone knowledgeable aboard who wasn't part of the official *El Faro* command chain. He could speak more freely and could either reassure her that they would be okay or maybe confront the captain if he thought they were in trouble.

In her usual light and girlish tone, she gestured to the BVS graphic: "Do you wanna see the pretty pictures with the pretty-pretty colors?"

When he saw the deep scarlets, intense blues, and bright saffrons on a normally pastel-hued map, Jeff swallowed hard. "Wow, look at that."

"That's us," she said, pointing to their current location. "And that's the storm," she said. "This is tomorrow." She clicked forward in time. The ship and the hurricane were destined to collide.

"Where are we right now?" Jeff asked.

She pointed to a spot forty miles northeast of Grand Bahama island.

"Really?" Jeff said, incredulous that they were heading straight for a storm system.

"Tentative," she said. "We are taking a track line a little bit further south, down here, so this is where we're gonna be in the morning," she said, confirming the direct hit.

Jeff spent some time studying the chart for possible escape routes. He saw a few deepwater channels that would get them out of the way. "Can you duck down between the islands here?"

Danielle knew Davidson. He didn't have any interest in changing the plan and she told Jeff as much, silently gauging his reaction. Full steam ahead to San Juan.

"Is the captain gonna grab Old Bahama Channel back?" Jeff asked, thinking about how much ocean Joaquin would churn up by Friday night.

"Yeah," Danielle said. Joaquin was forecast to intensify in a few days, so Davidson wanted to play it safe and sail on the lee of the islands, then skirt the Florida coast all the way back up to Jacksonville. "I think the *El Yunque* will be taking that route coming back up, too," she said.

In fact, Davidson had exchanged emails with the captain of *El Yunque* earlier that day and learned that his ship was sprinting back to Florida along their usual route. As they passed Joaquin, the captain informed him, they experienced gusts in excess of 100 knots. That on-site observation wasn't consistent with the NHC reports. Computer models and satellite imaging were already failing to accurately capture the rapidly intensifying storm.

What *El Faro*'s officers needed was a working anemometer, but the one onboard hadn't worked for some months. Several people had mentioned the problem to TOTE, but a repair or replacement was slow in coming. The ship's anemometer was so dodgy that Danielle joked to Jeff, "We'll just stick Jackie Jones out there. We'll use the *Jackie Gauge*. If he gets blown off the bridge, we'll say, *Oh, it was about 190 miles per hour*."

Understanding true wind speed and direction can be difficult when you're moving because you're creating your own wind. (Think of riding a bicycle on a calm day. Go fast and you'll feel a breeze. It's not the wind; it's you.) An anemometer uses vectors to make the calculation, taking into account the ship's heading and speed. For example, if an 18-knot wind is at your back and you're traveling at 18 knots, the apparent wind—the wind you perceive—would be zero, because you're both moving at the same rate in the same heading. Your vectors cancel each other out.

The anemometer is a critical piece of equipment because true wind direction can tell an experienced mariner where he or she is relative to a major weather system. A quick way to do this can be found in Nathaniel Bowditch's *The American Practical Navigator*—a thick navy blue book found aboard nearly every floating vessel. Continuously published since 1802, it's considered the mariner's bible. Bowditch describes Buys Ballot's Law, a handy rule of thumb invented by a Dutch meteorologist in 1857 that's based on the unassailable fact that in the Northern Hemisphere, hurricane winds blow in a counterclockwise direction. The rule says this: When your back is to the wind, stick out your arms, making a T. Your left hand will point to the low, your right to the high. The low, of course, is the hurricane's center. According to that rule, as long as winds were slamming *El Faro*'s port side, the crew could safely assume that the hurricane was dead ahead.

In the dark in a hurricane, however, it's all but impossible to determine true wind direction without an anemometer because the frenzied waves and winds create a chaotic condition, like being inside a Dyson vacuum. From the outside, the larger system makes sense. But when you're in the middle of it, it's impossible to organize the powerful forces into a clear, coherent picture.

Jeff considered their situation, but as an extra hand aboard the ship, he couldn't do much about it. Instead, he worried about how the weather would affect his weekend plans. He was rushing to finish his maze in time for the Halloween rush, and Joaquin could put a kink in his progress.

"The hurricane is going to shoot north after this?" Jeff asked.

"They're predicting that it will hit New York this Saturday," Danielle answered. That meant it would probably begin dumping rain on coastal Massachusetts by Saturday night. Jeff might get in a day of work.

Jackie Jones brought their attention back to the ship. He reminded Jeff that the Polish contractors needed to tie up their equipment.

"Absolutely," Jeff said. "The bottles are secured, pipes are all lashed down. So yeah, that should be all right."

"Lash down your workers?" Danielle joked.

"They're all excited," he answered.

"I don't think they realize what they're getting into," she said with a nervous giggle.

"See what color their heads are tomorrow," Jackie cracked, and the three had a good laugh. They knew they'd all be green tomorrow, too, if the weather turned as foul as expected.

At 2:30, Davidson came up to the bridge to remind Danielle that he was thinking about taking the Old Bahama Channel on the northbound run back to Jacksonville. Joaquin was forecasted to intensify in a few days, so Davidson wanted to play it safe and sail in the lee of the islands, then hug the Florida coast all the way back up to Jacksonville. He was obsessing about this proposed route change and was anxious to get TOTE's response to his request. Maybe he thought they'd thank him for being so proactive and conscientious. He was working hard to please the shoreside folks.

"I have to wait for confirmation from the office, but I put it out there," he said about the rerouting. "And I'll let you know."

"Does the company want to give permission now?" Danielle asked, half surprised, half exasperated. "Because it used to be just, *We're doing it. You people are sitting in your office behind a desk and we're out here. We're doing it. Hell yeah.*"

"Well, I'm extending that professional courtesy because it does add 160 nautical miles to the distance," said Davidson defensively.

"Yeah, but rerouting also saves stress on the ship," she said.

Why was she giving him a hard time? "That's why I just said, 'Hey, I would like to take this going northbound. I'll wait for your reply.' I don't think they'll say no. I gave them a good reason why, because if you follow this hurricane track down, then look what it does on October third, fourth, and fifth. And that's right where we're going."

"Yeah, and lightly loaded, it gets even worse," said Danielle. After unloading in San Juan, *El Faro* took on mostly empty containers. Without cargo weight, she sat higher in the water and would bounce around in the wind and waves.

"So I just put it out there and we'll see what happens."

Davidson didn't like his second mate's nervous energy. It rubbed him the wrong way. He had to put this whole thing to rest. "We're gonna be far enough south that we're not gonna hit the damn thing," he declared. "Watch. Gonna get a little rougher, but these ships can take it."

This exchange is one of the most contentious pieces of *El Faro* saga. Much time has been spent parsing Davidson's apparent request to take the Old Bahama Channel home, which he did via email earlier that morning to multiple managers at TOTE's Jacksonville office. Did he actually need the company's permission to reroute? His request was carefully worded and thoughtful:

"I would like to transit the old Bahama Channel on our return North bound leg to Jacksonville? This route adds approximately 160 nautical miles to the route for a total of 1,261 nautical miles. We need to make around 21 knots for the scheduled 10:05, 10:45 arrival time at Jacksonville pilot station.

I have monitored Hurricane Joaquin tracking erratically for the better part of a week. Sometime after 9/30 0200 she began a southwesterly track early this morning. Adjusted our direct normal route in a more southeasterly direction towards San Juan, Puerto Rico, which will put us 65 plus/minus nautical miles south of the eye. Joaquin appears to be tracking now as forecasted and I anticipate us being on the backside of her by 10/01 0800.

Presently, conditions are favorable and we're making good speed. All departments have been duly notified as before. I have indicated a later than normal arrival time in San Juan, Puerto Rico. Anticipating some loss in speed throughout the night. I will update the ETA tomorrow morning during our regular pre-arrival report to San Juan Port, etc."

Was Davidson planning to plow through in spite of the looming hurricane because he felt pressure to get to Puerto Rico on time? Had TOTE chastised him for playing it safe and going through the Old Bahama Channel in August during Tropical Storm Erika?

TOTE maintains that scheduling was never an issue. The captain is in complete command of the ship and its route. TOTE says that in all instances, safety comes first. The company claims that this email exchange was simply a formality—regardless of how one might read it, Davidson had full command of his vessel and his course. He did not need permission to take a different route.

That may be true at TOTE, but most ship's masters I interviewed tell a different story. Although a few can't believe that a company would pressure anyone to go into a storm, most captains (both foreign and domestic) say that fending off schedulers and managers is simply part of the job. The office worries about customers and profits; captains worry about everything else. Sometimes their interests diverge.

A captain's best attitude, my sources say, is the one that keeps crew and cargo safe. With pride, they've told their schedulers: *You can find another captain or fire me, but I'm not putting this ship or cargo in danger. I can always find another job.* They say that it's absurd to take orders from a person sitting behind a desk. He or she can't see what's going on at sea, usually lacks nautical experience, and can't imagine the conditions the crew is dealing with. Sure, the ship might be able to pull through a weather system, but if a new car breaks loose and smashes up fifty others, would it really matter that they arrived on time?

Regardless, some people, for whatever reason, cave to economic pressures. They assess their risk and decide they've got too much to lose if they defy the shipping company. It's not worth fighting. One ship's master told me about the time he was ready to ship out of Antwerp when a storm kicked up off the coast of England. He kept his vessel in port while another captain in his fleet chose to steam ahead. The ship that sailed was battered and beaten by the waves and winds, but it did get to the next port. The first captain showed me a photo on his iPhone of the resultant damage: mayhem aboard. Everything—the cars, trucks, and containers—was smashed to bits.

Davidson didn't get an answer to his request right away. On September 30, the one person on land in charge of keeping TOTE's ships safe at sea was attending an industry conference in Atlanta. His name was Captain John Lawrence, TOTE's designated person ashore (DPA) and the manager of safety and operations. Lawrence was responsible for the welfare of twenty-six vessels. When he was out of the office, it wasn't clear who was supposed to take calls or emails in his stead. TOTE didn't have a clear organizational chart to address that inevitability.

TOTE's officers aboard the ships didn't understand the company's structure, either. Davidson sent his message to half a dozen people, assuming someone at TOTE would eventually respond.

Lawrence seemed like a questionable choice for the job of safety officer at a shipping company. In 2010, he held a similar position at a New York–based tug and barge company when one of his Delaware River tug drivers pushed a barge right over a disabled duck boat filled with tourists, killing two people. The duck boat operator tried to radio a warning to the tug driver that he was heading straight for them, but never got a response. During the National Transportation Safety Board (NTSB) investigation of that incident, the tug driver admitted that he'd been using his cell phone and his laptop for personal reasons belowdecks instead of keeping watch. Following a criminal trial, the tug driver went to jail. What kind of safety culture had Lawrence fostered at that company that would allow such a thing to happen?

And now, as *El Faro* steamed full speed on a collision course with a hurricane, Lawrence was out of communication. No one in TOTE's office was tracking any of their vessels, nor were they tracking the developing storm. It's difficult to fathom how that could happen these days, especially since a ship's location is publicly available on any number of vessel tracking websites. You don't even need to log in. Just google a ship's name and you'll instantly find it anywhere in the world.

A logistics company's most valuable assets, along with the cargo and the people aboard, are its vessels. Why wasn't anyone at TOTE following them? One shipping company operator I spoke to said he considers the ships his children. He follows them constantly as they motor around the world. Denmark-based Maersk, among the world's biggest shippers, doesn't just follow its vessels' tracks. It also installs cameras and microphones in the engine room, on the bridge, and in the cargo holds, which send video and audio back to the main office in real time. This information can help the Maersk managers quickly identify, solve, and even prevent problems. TOTE had no such monitoring equipment aboard *El Faro*.

TOTE's reply to Davidson's request finally came in more than five hours later—a simple, "Captain Mike, diversion request through Old Bahama Channel understood and authorized. Thank you for the heads up. Kind regards." The email came from Jim Fisker-Andersen, director of ship management for TOTE, who was traveling back to his Jacksonville office from San Francisco, after overseeing some issues with the new LNG ship.

Why did Fisker-Andersen decide to reply to Davidson's email? "Because there was an unanswered question that I didn't want to leave open ended. I didn't want the email to go unanswered," he later told investigators.

When he was asked where Captain Lawrence was during that time, he said, "I don't know."

Did anyone onshore at TOTE track ships, Fisker-Andersen was asked. "No," he said.

At least one person was tracking *El Faro*, though: Second Mate Charlie Baird.

From his South Portland home, he was glued to the Weather Channel that day. And with his girlfriend's help, he was able to follow *El Faro* via her AIS (automatic identification signal) over the web. He watched in horror as she continued her course straight toward Hurricane Joaquin. Charlie sat all day on that sofa hoping against hope that someone on board had enough sense to turn the ship around.

CHAPTER 8

HULL NUMBER 670

Hull number 670 was born in Chester, Pennsylvania, on November 1, 1974. Christened *Puerto Rico*, it would be one of the last ships to be built at Sun Ship.

Don't bother trying to find the Sun Shipbuilding and Dry Dock shipyard. There's nothing left of the towering steel frames that once cradled ships while they were under construction, or the steel sheds that sheltered the pipe shop, the blacksmith shop, carpenter shop, and sheet metal shop. Nearly everything was leveled in 2004 to make way for Harrah's Philadelphia Casino and Racetrack. Where thousands of people built ships in a frenzy that lasted half a century, there's now a five-eighths-mile elliptical track for thoroughbreds to race around a neatly kept lawn. You can bet on the horses or lose your savings at the blackjack tables inside, but you probably won't be able to make out the skeletal remains of the towering shipways, now cut down to nubs, sulking at the water's edge across the track from the casino.

Pull over to the grassy shoulder along Route 291 outside of the casino and you'll notice a small cast-iron historical marker erected in 2007. It reads: "During WWII, Sun was the largest single shipyard in the world, with over 35,000 employees. It introduced the all-welded ship, which significantly increased ship production, and the T-2 oil tanker. Sun built over 250 WWII tankers, 40% of those built in the world and repaired over 1,500 war-damaged ships. Established by the Pew family, it was located at this site from 1916 to 1982."

In the summer of 1961, twenty-year-old John Glanfield was looking for work. Tall, slim, with a wide-open face and a meaty handshake, he'd graduated from high school a few years earlier and, being mechanically minded, was thinking about fixing cars for a living. Most of his classmates had left town to serve in the armed forces, but John felt compelled to stay close to home outside of Philadelphia and help his mother.

The Glanfields hadn't had it easy. John's father, orphaned and raised by German nuns in New Jersey, had served in Europe during World War II where he suffered a chemical weapon attack. He spent the next seven years in a hospital, lost a lung, and never fully recovered. He died the year John graduated from high school.

Growing up in Collingdale, Pennsylvania, John was bound to discover shipbuilding sooner or later. Fewer than three miles away on the banks of the Delaware River—the same river that General George Washington had crossed during an icy winter to surprise the British and win the Battle of Trenton—was Sun Shipbuilding and Dry Dock Company, the country's largest shipyard. One day, John hopped on his bike and pedaled out to the Delaware River, where he was immediately offered a job at the busy shipyard. The pay wasn't great, but the future opened wide.

Sun Ship was a tight, mature operation at a moment of intense innovation. As soon as he saw the gigantic steel hulls under construction, and the four thousand people employed to paint, weld, rivet, and outfit them, John was hooked. "I can remember my first day walking up the gangplank onto a ship that was only partially built," he told me. "It had the double bottom part of the side shell and there were all these guys in the engine room and all these planks and staging. And it was like, oh my God, I just love this. I was in love with the thing right off the bat."

For a twentysomething guy with nothing but a high school degree, Sun Ship offered the opportunity to be part of something big, a chance to grow. John apprenticed as a pipe fitter and worked his way up, taking small salary bumps as he went, earning a few more dollars each year for his young family, until he was overseeing the work of dozens of men. Sun Ship took care of its supervisors, providing a complimentary hot lunch to all managers like him. On Thanksgiving, the company gave every single worker a turkey and $25.

Sun Ship's own history followed the victories and setbacks of America itself. It was founded during World War I by two brothers, J. Howard and Joseph N. Pew, now better known as the progenitors of the Pew Charitable Trust. The brothers had inherited the seeds of a great fortune from their father, Joseph Newton Pew, who had bet big on the industrialization of a nation emerging from the ravages of the Civil War.

In the late nineteenth century, Washington, DC, finally unfettered by the demands of the agrarian south, financed the railroads, canals, and skyscrapers that would drive a new economy. Joseph Newton Pew invested in the lifeblood of industry: energy. He bought up Texan oil fields and tapped into natural gas veins in the backwoods of his native Pennsylvania. Like Andrew Carnegie and John D. Rockefeller (who'd eventually do business with him), he profited not just from government investment but also from the seemingly endless waves of immigrants pouring in from Europe and Asia who were willing to work for little pay and even fewer labor protections.

Joseph Newton Pew died in 1912, leaving Sun Oil Company

(later Sunoco), in the hands of his highly educated sons. By 1916, the pair agreed to take over their entire energy distribution chain by building the hulking tankers necessary to transport their liquid gold from their oil fields in Galveston, Texas, to their refineries near Philadelphia, and then around world.

On October 30, 1917, a few months after the United States entered World War I, Sun Ship's first vessel, a 430-foot tanker christened *Chester Sun*, was launched off the banks of the Delaware River. "In spite of the terrific downpour," J. Howard Pew told the *Philadelphia Inquirer* the next day, "Everything went as smoothly as clockwork. We wouldn't postpone the launching, for the superstition of the sea has it that if a launching is postponed, the ship will be a 'hoodoo' vessel [a cursed or bad-luck ship]. If everything goes as smoothly for the Chester Sun as the launching went, she should be anything but a 'hoodoo.'"

Dozens of orders kept the yard humming for years, especially from the company's oil division, which drove Sun Ship's engineers to constantly innovate their designs. They perfected the welded hull—lighter, and quicker to build than the riveted hull—and the double hull, which added a layer of protection against leaking.

When the Depression hit, orders for new ships slowed to a trickle. But the Pew family, like many major owners of the day, believed that companies had an obligation to take care of their employees. Sun Ship retained nearly all its workers by paying them two days a week, whether or not there was a ship in the yard.

Keeping employees through tough times paid off for Sun Ship. By 1937, as World War II loomed, the US government began looking for a shipyard to fill orders for hundreds of vessels. Sun Ship was ready to go, replete with a huge, trained staff. At its peak during the war, the shipyard employed more than thirty-five thousand women and men, including my paternal grandfather, who had taken a hiatus from his clothing manufacturing company to sire my father and join the war effort. In 1944 alone, Sun Ship launched an astounding seventy-five tankers and military vessels.

When John Glanfield rode his bike to the shipyard in 1961, the place was abuzz, but with a different kind of energy. Sun still built ships, about six or so per year, but it had diversified. Taking advantage of its sophisticated large-scale industrial design, engineering, and custom machining departments, its executives began landing contracts in a broad range of industries. Working inside the cavernous hull of a seven-hundred-foot-long steel ship was exciting enough, but the occasional top-secret government projects that came along made a young guy like John thrill with excitement.

John built rocket casings for NASA, nuclear reactors, catalytic towers for the oil industry, wind tunnels, and—the one that he still rhapsodizes about in hushed tones fifty years later—the *Hughes Glomar Explorer*, a decoy deepwater drilling rig financed by the CIA.

The official story was that billionaire Howard Hughes commissioned the vessel for his underwater mining company. In fact, "Project Jennifer" was paid for by the CIA to recover a sunken Soviet submarine lost off the coast of Hawaii. It's rumored that the sub was sunk when a rogue team of Soviet sailors wrested control and tried to fire a missile at Hawaii, tripping a fail-safe, which caused the missile to explode as it was exiting the torpedo tube. The agency desperately wanted to study the sub to prove that America's Cold War opponent had equipped its vessel with nuclear missiles and torpedoes.

While the Soviets hunted in vain for their lost vessel, Sun Ship's offshore drilling decoy parked right over the sub, precisely located by the navy's system of hydrophones in the area, which had heard the doomed vessel explode and sink. Once in position, the belly of *Hughes Glomar* yawned wide above the ocean floor, dropped a giant claw thousands of feet down, clamped onto the sunken sub, and pulled the mangled wreck up into her hollow hull. After closing up, the operators dewatered the hold and stood in awe of their prize.

Shipbuilding was demanding work but John was enthralled. He loved the "camaraderie of the people, the fact that your work was appreciated," he says. "Guys worked hard in all kinds of weather—rain, snow. The heat in the summertime was brutal. In the hull of a ship with the sun beating down, sometimes the tools got so hot you couldn't pick them up."

He'd come home dirty, tired, and satisfied, his clothes riddled with burn holes from errant welder's torches or tangles with sharp objects. His wife, Dot, learned to shop in thrift stores, buying John three used coats per season; after a few months at the shipyard, they'd be nothing more than rags. Each night, John would stomp his shoes, knocking off the white, glistening, deadly flakes of asbestos. The heat-insulating material, cut by handsaws on-site, was used to wrap the steam pipes of the turbine engines that drove ships' propellers. John would shake out his jacket before sitting down to dinner, the kids playing nearby. "I'm lucky now," he tells me, "because all the people that I worked with are dead. The difference is I never smoked. If you smoked, you were doomed." Researchers have determined that smoking can increase risk of asbestos-related diseases by 90 percent.

John learned how to sequence construction in the shipways, very useful knowledge for the elite naval architects working at their desks in Sun's brick office building, worlds away from the noise and chaos of the construction zone. Even though he lacked an engineering degree, he was frequently pulled into meetings to help the architects work out the complex steps required to execute their designs.

One day, hull designer Eugene Schorsch presented the Sun Ship team with plans for a sleek, revolutionary new class of vessel he was working on, something called a roll on/roll off "trailer ship." This one would be a game changer.

At the time, the United States ruled the global economy. Because Europe hadn't yet fully recovered from the ravages of war, American exports were in high demand. A sign on a bridge up the Delaware River installed in 1935 proudly declared: TRENTON MAKES, THE WORLD TAKES. It was never more true than in the early '60s. America had laid the groundwork for industrialization after the Civil War. Now it was reaping the rewards, shipping its goods everywhere, especially to Europe and Asia.

But imports were beginning to creep in. By 1970, the United States would experience a trade reversal from which it would never recover.

To stay competitive, American manufacturers had to up their logistics game. After all, most of their exports had to traverse at least one ocean—either the Pacific or the Atlantic. Unless someone figured out how to ship goods cheaper and faster, Americans would lose market share to local outfits in Asia and Europe. But loading and unloading cargo was still stuck in the nineteenth century. On the docks, boxes, bags, and goods of all sizes and weights were at the mercy of an army of longshoremen who worked with ships' crews to strategize how to pack all that random stuff into a hull.

Sometimes it would take days to load a vessel, affording seamen plenty of time to blow their money on drink, girls, or gambling. That was, of course, part of the fun of pursuing a living on the high seas, but it wasn't efficient. And it was expensive. The more handling, the more expensive the export.

As long as a ship idled in port, it was losing money. But nearly every American port depended on unionized longshoremen and stevedores to carry cargo onto ships, then stow and lash it down. These thousands of dockworkers were pricey and potentially problematic. If they called a strike, they could close the port, costing shipping companies serious money. They'd done it before; they could do it again.

American manufacturers couldn't risk losing their customers abroad. Any interruption in the supply chain presented yet another risk in an inherently dicey business. Economics were one thing. Political turmoil was another. Weather always loomed large, ready to thwart a shipper's best-laid plans. But operating at the mercy of organized labor was an insulting vulnerability. How could shippers cut out the uncertainty of people from the logistics equation?

The answer was containerization. A one-size-fits-all shipping crate that could be packed, shipped, and reused would make things go a lot faster. Regardless of what you loaded in the box—rubber duckies, underwear, guns—it would stack neatly onto a ship's deck along with the others, just like children's blocks. You'd need a lot fewer dockworkers. No more thinking. Just pack, stack, and go.

You'd have to standardize a lot more than the shipping crate to get containerization to work. Successful implementation would require every port of call to embrace a universal system because only special cranes could load and unload such big, heavy crates. Someone had to patent an open-source design for the box itself. Every naval architect would need to imagine new vessels to accommodate the box's dimensions. From New York to Papua New Guinea, all ports would have to invest in container-ready cranes to accommodate a standardized mounting clip system, and clear huge open lots nearby to warehouse empty boxes. Someone would have to design and build new flatbed trucks onto which containers could be lowered directly from the ship and driven straight onto the interstate.

Perhaps the biggest obstacle of all: shipping companies would have to convince the longshoremen's unions to accept their sad fate. No doubt, containerization would destroy thousands of dockworkers' jobs.

For these reasons—the considerable outlay of cash coupled with insurmountable labor resistance—containerization took a while to catch on.

But you couldn't fight progress for long. In the mid-'50s,

North Carolina–based trucking tycoon Malcom McLean developed a complete containerization system, which he launched with a single experimental ship in 1957. A year later, his company inaugurated containerized service between the United States and San Juan, Puerto Rico. Over the next decade, all eyes were on McLean's SeaLand, the revolutionary shipping company that had bet on containerization. A few ports adopted McLean's system, but others watched and waited.

Meanwhile, Sun Ship made a different kind of bet.

In 1965, a groundbreaking ship order landed on Sun Ship hull designer Eugene Schorsch's desk. Unlike the "deadly detailed specifications heretofore commonplace," he says, this order was just one and a half pages.

Schorsch had worked at Sun since 1952. He'd taken a job there straight after graduating from the premier naval architecture school in the country—New York's Webb Institute, which, to this day, offers free tuition to all students. Taking a walk around the yard in 1952, Schorsch observed, "It was rugged. It was busy. Perhaps six thousand people working on a dozen ships. Some streets were paved with old riveted shell plates to make them less muddy. Philadelphia still had horse-drawn wagons and the horses had not long left the shipyard."

Like Glanfield, Schorsch knew he'd found his calling. "It looked like the kind of place where you could learn to be a shipbuilder," he wrote in a trade journal many years later, "where the ghosts of thirty-six thousand people and twenty-eight shipways offered a young person a place to hone his book learning and get to know practical, real skills."

In the thirteen years since he started working for Sun Ship, Schorsch had designed all kinds of things, but this would be the first time his team got to design a revolutionary ship from scratch.

Sun Ship was itching to get into the lucrative cargo shipping business and this new vessel was slated to be the first of a fleet of trailer ships for the Transamerican Trailer Transport (TTT), set up to compete with SeaLand. Maybe containerization would take off. Maybe not. This was a hedge.

Similar to McLean's container ships, Schorsch's merchant vessel would be built for the age of the interstate. Replacing the laborious loading and unloading methods of the past, all cargo would be driven or towed aboard the ship over ramps and discharged the same way in less than twelve hours. That meant that the ship's interior had to be free of obstructions, so that the 260 trucks loaded with goods could drive straight from the dock via a ramp onto the weather deck, where their trailers were unhitched from their cabs and secured to the ship. Another three hundred cars and light trucks would be driven from the dock and down into its cargo holds via internal ramps. *Roll on.*

The ship had to have lock boxes on deck where the trailers could be quickly clamped in place and plenty of D-rings for lashing down other vehicles. Once at their destination, the trailers and cars had to be unlocked just as fast so that a stevedore could drive them right out again and onto the highway. *Roll off.*

The ship Schorsch was designing also had to be fast. In the increasingly competitive market that was global shipping, speed was everything. The vessel was slated to do a weekly turnaround between New York and San Juan, Puerto Rico, requiring normal operating speeds in excess of 23 knots. "Maintaining a regular schedule was an important selling point to shippers," a report of the day announced, "since the plan was for New York cargo to be loaded on Fridays with a Friday evening departure for San Juan, putting the ship in San Juan before 8:00 a.m. each Monday morning." So she was built narrow and powerful, with a thirty-three-thousand-horsepower steam turbine engine that spun a single giant propeller.

The roll-on/roll-off trailer ship (ro/ro) presented other unique engineering challenges for Schorsch's team as well. While the ship was being loaded, the weight of the tractor trailers driving over that huge steel ramp onto the second deck would pull her over, so Schorsch developed a counterweight system of ballast tanks and pumps that could keep the boat stable under dynamic loads. He also had to design the ship's framing robust enough to support a wide open interior; from fore to aft, the ship lacked any vertical structure along its centerline. If a hold flooded, water would slosh from port to starboard as if it were a giant swimming pool, throwing off the vessel's roll.

The weather deck, complete with its huge doorway cut into the hull, was a key feature of the ro/ro class. It allowed the dock-to-ship ramp to remain nearly level during loading. But that, combined with its massive, undivided cargo spaces, would prove the ro/ro class's greatest vulnerability.

In the ensuing years, the United Nations' International Maritime Organization (IMO) has revisited this class many times, questioning the inherent instability of ro/ro design and exploring ways to make it safer. While ro/ros statistically incur as many accidents as other ships on the high seas, they have caused a disproportionate number of fatalities. In part, that's because passenger/car ferries are considered a part of the ro/ro class. The most horrifying accident involving one of these ships occurred in 1994, when the passenger ship *Estonia* suddenly sank in a terrific storm in the north Baltic Sea, taking nine hundred people with it. Authorities soon determined that the ship's outer bow door had been torn off during the storm, allowing water to rush into the expansive car deck which caused the vessel to list, roll over, and sink very quickly. Cargo ro/ros had the same weaknesses—their huge vehicle decks made them susceptible to rapid, catastrophic flooding.

Another nagging problem for this class of ships was its lack of stability caused by cargo movement. In the cavernous holds, one loose trailer could inflict havoc. The third point of vulnerability was the ro/ro's low freeboard—doors and other openings in the hull deliberately built close to the waterline for easy on, easy off. In good weather, these weren't a problem. But in a storm, or if shifting cargo caused a sudden list, a lot of water could rush onto the deck; if any watertight hatches were left open, that water would pour into the hold and quickly swamp the vessel.

By the mid-'70s, all these issues were known by organizations set up to monitor shipping, but demand for ro/ro cargo ships and ferries far outpaced safety concerns. It seems the message didn't get down to those hired to man these ships. None of the experienced mariners I spoke to had any awareness that the ro/ro was considered more vulnerable than any other.

Schorsch's first roll-on/roll-off ship was seven-hundred-foot-long *Ponce de Leon*, launched from Sun Ship in November 1967. *Ponce de Leon* is an important part of this story. She was the older sister of *El Faro*. The two vessels shared the same DNA and were built in the same yard by the same shipbuilders, one of whom was John Glanfield.

Both ships were designed to compete with SeaLand for the Puerto Rican trade. In fact, *El Faro*'s first name was *Puerto Rico*.

When she was built, *Ponce de Leon*'s design was so groundbreaking that Schorsch and his colleagues presented a paper at the American Institute of Aeronautics and Astronautics' 2nd Advanced Vehicles and Propulsion Meeting in Seattle, Washington, in May 1969, describing the ship's features. The paper emphasized the fact that creating *Ponce de Leon* involved designing an entire *system*, foreshadowing containerization:

> "The modern concept of a ship is to view it as part of an ocean spanning total transportation system. The PONCE DE LEON, more than most ships, can be thought of as part of such a system. In fact the story of the design and construction of the PONCE DE LEON is the story of the creation of the system of which it is a part. The PONCE

DE LEON introduced Roll-On/Roll-Off shipping to the New York/San Juan trade, and in fact to the world, on a larger scale than has heretofore been known, in the sense that it is the largest Roll-On/Roll-Off ship in operation.

"Although the ship to be used in the system was the largest single item of capital expenditure involved, shore based loading ramps had to be designed, built and shipped to the two ports. A fleet of special loading and unloading tractors had to be provided at each port. Trailers had to be constructed and an operating organization established. Establishing working arrangements with the appropriate labor organizations and training tractor drivers for the driving conditions on the ship was no small part of the task."

John Glanfield and his fellow shipbuilders admired the all-white *Ponce de Leon*'s sharp, narrow prow and long, elegant sheer— the term that describes the way the main deck line peaked at the bow, dipped low midships, and rose again at the stern. She had that classic look of a '67 Stingray—sharp, edgy, built for speed. Inside, her two boilers powered a steam-propelled turbine that spun her single screw through the water at impressive speeds. The *Ponce*-class ships, all ten of them built between 1967 and 1977, served their crews well, and lasted an average of thirty-six years before being scrapped. As of this writing, two are still being used. And so began the life of an American-flagged, American-built ro/ro vessel, constructed and maintained so well that she ran decades longer than many of her foreign-flagged counterparts.

As planned, the SS *Puerto Rico* faithfully delivered goods to and from Puerto Rico for fifteen years, first for TTT, and then, when the commonwealth nationalized its shipping, for Navieras de Puerto Rico Steamship Company, which had purchased three of her sister ro/ro ships as well as part of the deal. In the meantime, the Sun Company introduced its own shipping service called Totem Ocean Trailer Express (TOTE), running cargo with its fine ro/ro ships between Tacoma, Washington, and Anchorage, Alaska. Sun's ownership of TOTE didn't last long. A consortium of businessmen based in Seattle, now known as Saltchuk, bought TOTE in 1982.

In 1993, TOTE bought back *Puerto Rico* from Navieras and had her lengthened ninety feet at an Alabama shipyard. Many of the other *Ponce*-class ships had been lengthened too. In fact, even while it was building its ro/ro vessels, Sun Ship noted that they could be lengthened and did some of that work on other hulls at the Chester shipyard before it closed.

Because the coast guard considered lengthening a major conversion, the new nincty-foot section of *Puerto Rico* had to meet current codes. The rest of the ship remained intact, including *Puerto Rico*'s two gravity-launched "open" lifeboats. International codes required all ships built after 1986 to have enclosed lifeboats; the watertight vessels look like small, bulbous submarines. But American shippers balked at the cost of purchasing and installing these pricey add-ons and the US Coast Guard was generally sympathetic to industry. And so, without much resistance from regulators, *Puerto Rico*'s lifeboats, banned everywhere else, were grandfathered in and remained on the ship until the day she went down.

Saltchuk, through TOTE, continued to expand its holdings, acquiring Sea Star lines in 1998, digging deeper into the Puerto Rican trade route. By now, there was a lot of competition in that market. Someone was going to lose his shirt.

When *Puerto Rico* was lengthened, she was renamed *Northern Lights* and sent into the Pacific Northwest trade. During the second Persian Gulf War, she was chartered by the US government to carry military cargo to Iraq.

Of course, by 2003, McLean's containerization had conquered the world. Yes, the ro/ro ships still had their uses, but importers and exporters demanded container capabilities of their shippers as well. Speedy old *Northern Lights* would need to be reconfigured once more, this time to take dozens of containers on her main deck. And again the coast guard gave in, allowing her to sail with her white fiberglass open lifeboats rather than the bright red submarine-like enclosed ones found on most other modern ships. In a wild storm, one could quickly sink or capsize while the other would get knocked around, but stay afloat.

Sun Ship discontinued new shipbuilding on January 9, 1981, the victim of appalling mismanagement in the preceding years and a shady sale to an entity Schorsch is convinced had some connection to the CIA. Enter the go-go '80s, trickle-down economics, and the golden age of management consulting and offshoring. After the ownership change, executives determined that the land Sun Ship occupied was more valuable than the ability to build and repair ships. The whole factory shut down, and a year later, its parts were sold for cash.

Sun Ship's sixty-year story speaks to America's great maritime boom and bust. Its hulking sheds and dry docks decaying along the banks of the Delaware River echoed the postindustrial landscape that began dominating America's heartland—Gary, Indiana; Detroit; Pittsburgh; Cleveland—as early as the 1970s. By the time Reagan assumed power, the country's shipbuilding days were over, timed perfectly to coincide with a new imports-welcome administration.

America is now number one, in trade imbalance. Since 1975 the United States has maintained the largest trade deficit in the world, nearly four times the total of number two, the UK. Both were once maritime superpowers. China has the greatest trade surplus. It also dominates the global shipping industry.

The Philadelphia-built *Ponce*-class ships traveled through the

Panama Canal, sailed in the rough waters off Alaska, and delivered military goods to the Middle East during the two Gulf Wars. They survived long past their shelf life. Few questioned their seaworthiness because no one wanted to pay to replace them.

Like all things under the sun, the vessels had vulnerabilities. Only one of them was pushed to the breaking point.

CHAPTER 9

AFTERNOON

27.04°N -77.12°W

Jackie Jones stood at the ship's wheel, keenly scanning the ocean for signs of Joaquin. "Looks like the wind's starting to pick up a little bit."

"Yeah, a few more whitecaps out there," Danielle said distractedly.

She was mentally exhausted.

She'd watched the hardworking mariners she respected get screwed by TOTE. She saw captains and chief mates get fired. She saw officers getting promoted, seemingly at random, to the new ships while she languished on an old vessel, bound to get scrapped. She saw the ship itself deteriorating. Davidson was a newcomer, a neophyte, a poor substitute for the profoundly experienced men fifteen years his senior who'd protected her, looked out for her, trained her.

Danielle's mother says that her daughter had a strong sense of order that was shaped by her ex-navy parents who, after serving for years in the military, instilled in their children respect for discipline and authority. Add to that years of watching Fox News, and Danielle was confident that she knew right from wrong. She was a die-hard conservative who could tell you exactly where she stood on nearly every issue—politics, abortion, welfare, regulation. Following strict conservative tenets, all answers were as clear as day. She never shied away from a political debate.

Danielle's politics didn't square well with the fact that her job depended on a complex network of protectionist laws, strong unions, heavy regulation, and corporate tax breaks. It was the "small government, pro-business" crowd that had systematically undermined the coast guard's ability to properly regulate America's commercial fleet. (Notably, *El Faro* was scheduled to go on the coast guard's "watch list"—a file of troubled ships that needed extra oversight because they'd been in incidents, or because they were exceedingly old—on October 1.)

But these were complex issues, and Danielle preferred clear, concise commands. In that way, she made a good deck officer.

Because she shipped out so much, dating wasn't easy, but she'd recently begun spending more time with Chief Engineer Jim Robinson who was off-duty at his Maine home during this voyage. Robinson had been assigned to oversee *El Faro* when she went into dry dock that November. (Following US Coast Guard regulations, *El Faro* had to get a major overhaul every five years.) The Bahamas shipyard assigned to the task had been given a lengthy list of things that needed to get fixed before the vessel went to Alaska.

"Jimmy hates going to the shipyard," Danielle told Jackie. She thought that the company had taken advantage of Robinson during the last dry dock session, and she was encouraging him to ditch TOTE. "I keep telling him, 'Why the hell do you do that to yourself? Obviously, all that work's gotten you nowhere with this company.' They don't appreciate it. He was putting in twenty-four-hour days—literally twenty-four-hour days at the last shipyard. And he got nothing for it except for a *Fuck you—you burnout—we don't want you for the new ships.* He's an engineer so he won't have any trouble finding a new job at a company that actually appreciates hard work."

Her thoughts were interrupted by the arrival of Jeremie who had come back to the bridge with Davidson for the change of watch. Danielle took the third mate over to the chart room to update him on the storm. "Here's what it's looking like now. At about 23:00 we should be full force slamming into it." Then she added with a nervous giggle: "It's a good thing I'm the swell whisperer," implying that she was skilled at steering the ship in rough seas to get a smooth ride.

Just as they were wrapping up, an emergency message broadcast over the VHF radio from a US Coast Guard C-130 plane flying overhead:

"Sécurité. Sécurité. Sécurité. The National Hurricane Center has issued a hurricane warning for the central Bahamas, including Cat Island–Exuma–Long Island–Rum Cay–San Salvador. The National Hurricane Center has issued a hurricane watch for northwestern Bahamas, including the Abaco–the (Canary) Islands–Bimini–(Elliotbrook)–Grand Bahama Island and New Providence. The coast guard requests all mariners use extreme caution. The United States Coast Guard aircraft standing by on channel 16."

The coast guard sometimes sent a plane out to warn small crafts lacking *El Faro*'s sophisticated communications equipment about approaching hurricanes, but it was rare for the ship's crew to actually hear the message. They usually sailed too far out to pick up the signal. Besides, the past couple of years had been quiet, hurricane-wise.

"Aircraft he said?" Danielle asked, surprised that planes had come all the way out there to warn vessels in their vicinity. "Oh wow."

"Wow," Davidson repeated. "So by 2:00 tonight—on your watch—you should be south of this monster," he stated confidently.

Davidson assured her that he'd stay up with her, to help her get through. "I will be up the entire night for the most part, once we start getting into the shit," he said.

They'd try to keep the ship from rolling too much, he told her, but other than that, they'd stick to their course.

Danielle wasn't so sure. She reminded Davidson that AB Jack Jackson had sailed in Alaska in rough weather on *El Faro* and had told her that the ride was horrendous. "He had much experience with the weather with this particular ship," she told him.

But the captain didn't want to hear petty warnings from his second mate and brushed her off, saying he was going to go "watch a little television, I think."

When Davidson left them alone again, AB Jackie Jones didn't conceal his anxiety: "It's all whitecaps out there now. And the swell's growing."

The aircraft broadcast blasted over the radio once again: "The coast guard requests all mariners to use extreme caution."

A few minutes later, at 3:00 p.m., *El Yunque* appeared on their radar, thirty miles out, steaming northwest at 22 knots. *El Faro*'s sister ship was racing back to Jacksonville. "They're trying to get away from the storm too." Danielle said.

"Nobody in their right mind would be driving into it," Jackie said.

"We are," Danielle said, and then she laughed. Because what else could she do? She'd been put on the spot, in charge of steering a loaded, ancient steamship through an erratic hurricane. She was relatively new to the second mate job and was still learning the basics. She didn't have the experience or knowledge to improvise a course through deafening winds and roiling seas at night. That was someone else's responsibility. It was the captain's. Not hers.

As the clouds thickened, the satellite TV went in and out. Without the Weather Channel and the news, they had to rely solely on BVS and the text-only weather reports from the National Hurricane Center.

Danielle yawned. She hadn't had much sleep. Yesterday, while the ship was loading, she had spent a lot of time on the phone with the oil company in Maine trying to cancel her account. "I'm not paying for someone else's fuel," she told Jackie. "The cable company wants to rip me off too." Closing down her Rockland house so that her mother could sell it had been a headache. She'd need more caffeine to get through this watch and poured herself another cup of coffee.

26.39°N -76.50°W

A little before 4:00, Shultz took over the watch.

"Joaquin's still coming right for us," Danielle told him. "Or we're headed right for it. The swell has been about the same— about six to eight feet. From the east. But the wind has started to pick up more." She mentioned that the boatswain had tightened up the lashings on the trailers on the second deck in preparation for the storm.

A few minutes later, the chief mate of *El Yunque*, Kwesi Amoo, radioed to *El Faro* and bantered with Shultz. Talk was light, yet pointed. Amoo said that his ship had sped up to escape the storm.

"We're trying to give it an extra thirty to fifty miles from the predicted center as we scoot around here," Shultz replied.

"Eh, just go through it," joked *El Yunque*'s chief mate.

"Roger. Yeah, I've been there. No thanks. I just wanted to say hello and thank you very much. And as always, wishing you well. See you on the way back."

Before signing off, Amoo said, "The captain here is saying you're going the wrong way."

Shultz laughed. "No," he said. "We're really loving that BVS program right now." Thanks to BVS, he explained, he and Davidson knew exactly how the story would go. BVS even told Davidson the precise minute he would safely arrive in Puerto Rico, based on his heading and current speed. The BVS interface did not equivocate, suggest uncertainty, or offer a range of possibilities. In the world of BVS, the future was fixed and knowable. They'd be fine.

The master of *El Yunque*, Captain Kevin Stith, got on the radio. He and Shultz exchanged pleasantries. Then Stith brought up the other elephant in the room: lack of job security at TOTE.

"I never know where I'm going next," Stith said. "So we'll see what the company has in store for me." Another TOTE captain worrying about his job.

"Are you actually without a plan?" asked Shultz, a little surprised.

"Anything and everything could happen. They haven't told me anything, you know. Kinda like the way they left me down there in San Juan for three days. They just said, *Go to that ship*. Next thing I know, I'm hanging out at the beach all day. So anyway, the short answer is, no. I have no idea."

Shultz hung up the receiver and looked out at the ocean. Maybe he should start worrying about his job instead of the weather.

He scanned the latest printed NHC forecast. "It's pretty much what we knew this morning," he told his helmsman, Frank Hamm. "We're predicting 40-knot winds on our starboard beam tonight. The latest and greatest forecast. But that's not hurricane force."

According to Buys Ballot's Law, the only way they would get winds on their starboard beam while sailing south is if they got below the storm. Shultz was confident they would.

Frank Hamm had kept to himself when he was steering the ship during his early morning shift. Now he was wide awake, and worried. He was a big man, forty-nine years old, and diabetic. His thoughts wandered back to Jacksonville where his wife, Rochelle, and three children lived. He'd met his wife twenty-seven years before while working at a Red Lobster in DC. Her father worked in shipping and when he got transferred to Jacksonville, the young couple followed him down. One day Frank's next-door neighbor encouraged him to join him in the merchant marine. The life was good—just fill out the paperwork, pay the fee, take a test. Frank got his certification at the union's school in Piney Point, Maryland, and started shipping out in 1999.

Frank was gone for long stretches of time, but he and Rochelle made it work. For a while, he had a job in Charleston, South Carolina, and Rochelle would drive their car right onto the ship when Frank was docked for a romantic evening.

Between gigs, Frank would hang out at the Jacksonville union hall where a wall of monitors posted open jobs. Every company, every ship. As he worked his way up the ranks, he could take his pick of jobs—around the world or just a week. "He'd be at the hall all day," Rochelle remembers. Sometimes he'd grab a ship leaving port that day. The shipping company would send a car for him. He kept one bag packed full of warm weather gear, and one packed for cold climates. One time, a ship he was on changed course and sailed far north, catching him unprepared. He was so big, Rochelle says, "they had to sew together two jackets for him. I had to FedEx his meds."

Shipping took Frank places, and he always brought home gifts, though sometimes they were a little weird. From New Orleans, he brought Rochelle a voodoo doll. *Oh, Frank.* She shook her head and smiled when he gave it to her. *Why can't you bring me a damn T-shirt or a keychain?* "I'm a Christian and the devil is real," she said with a laugh.

At 4:30 in the afternoon, *El Faro* was passing just east of the Northeast Providence Channel—the only deep, wide trench that provided clear passage between Little Bahamas Bank and Great Bahamas Bank. If they wanted to escape to the Straits of Florida on the lee side of the islands, they'd have to practically turn the ship around to squeeze in there, but that would get them to the safe side via the Northwest Providence Channel. Their next escape route west would be the Crooked Island Passage, more than nine hours southeast of them. There'd been some discussion of the Providence Channel between the chief mate and Davidson early that morning, but it'd been dismissed, deemed too far out of their way.

"So this storm is gonna start around 'bout when we get off watch?" Frank asked Shultz.

"Yep," Shultz said, looking at Joaquin's projected track. He surmised that winds would swing around as the ship moved southeast of the storm. "And then 5:00 in the morning we should be settling down with about 35 knots of wind on the starboard side. If we'd stayed on our original track, we would have hit dead center."

Davidson came up to the bridge to tell Shultz that he'd requested the Old Bahama Channel route on the way back. From the captain's perspective, the hurricane looked like it would intensify in a couple of days. Joaquin was just warming up. He wanted to be far away when it did.

Frank Hamm overheard the two officers. "Is there a chance that we could turn around?" he asked. He knew he was at the bottom of the hierarchy, but at this point, he was getting tired of the officers' silence. He didn't like what he was seeing and had to speak up.

"Huh?" said Davidson, perturbed that the helmsman was getting involved. Officers and unlicensed seamen rarely discussed routing decisions aboard the ships.

"I'm just bein' nosy," Frank demurred.

"Oh," said Davidson, relaxing. "No, no," he stated definitively. "We're not gonna turn around." Then again, more emphatically: "We're not gonna turn around."

He went back to looking at the charts and spoke his thoughts out loud to reassure the men on the bridge that he'd thought everything through. "The storm is very unpredictable. Very un-predictable. It went high, it went left, it went right, it went back again. This one in particular is very erratic. So we'll get away from all that when we leave San Juan. I want to come home up Old Bahama Channel and not get tangled up with this thing."

Word spread around the ship that they were taking the Old Bahama Channel back, which lifted everyone's spirits, including Danielle's. She came up to the bridge a bit before five o'clock so that Davidson and Shultz could get dinner in the officers' mess.

"I love taking the Old Bahama Channel," Danielle told Frank before he went down to supper. "It's really pretty when we get close to the islands. You can see them out there, especially the Dominican Republic. It's a beautiful run."

Once the captain was gone, Danielle returned to the chart room to plot out the newest forecast from the NHC while talking it through with helmsman Jackie Jones. What she saw alarmed her.

"See where we are? Right here. Eleuthera Island." It is a long, narrow sandbar of an island east of Nassau. Then she read the latest alert: "'Government of the Bahamas has issued a hurricane warning. Northwestern Bahamas, including the Abacos, Berry Islands, Eleuthera, and Grand Bahama Island.'"

"That's awesome," she joked, trying to ease her fear with humor. "But all of this area is a hurricane warning. All of this area here: 'The government of the Bahamas has issued a tropical storm warning for the southeastern Bahamas including the Acklins, Crooked Island.'" She let out a big sigh. What the hell were they doing out there? "So at 2:00 in the morning, it should be right here," she said, referring to the forecasted point on her map. "Let's see where we will be."

She followed her line, then she couldn't help it. She laughed like hell. "We're gonna be right there with it. Looks like this storm's comin' right for us. You gotta be kidding me."

She plotted her points again, just to make sure. She preferred to plot the weather from the NHC rather than rely on BVS. No way she could be right, she thought to herself.

Joaquin was a compact, soon-to-be powerful storm with an indeterminate eye. Danielle couldn't have known, but even the NHC's positions of Joaquin were seriously off. For the next thirty-six hours, official reports put the eye as much as forty miles too far to the north. Lack of reliable data had hindered the NHC's reporting accuracy. The storm system had developed much faster than anyone expected, defying all the odds, intensifying at an astonishing rate. The NHC's Hurricane Hunters usually ran through a storm system collecting data every twelve hours; keeping track of this rapidly developing hurricane would have required much more frequent flies. The NHC's forecasts were also handicapped by the fact that there weren't any ships in the area sending voluntary weather reports. Without quality data, the meteorologists could only get a fuzzy picture of the storm from satellite imagery. There simply wasn't enough information to accurately predict Joaquin's intensity and path.

The BVS forecasts were even less accurate because they lagged several hours behind the NHC's forecasts. By the time the captain downloaded them, they were based on data more than twelve hours old. Here's why: NHC meteorologists use dropsondes, ship-reporting, and satellite data to collect raw data from land and sea. They run this mountain of data through dozens of computer models that then produce a range of possible hurricane paths and intensities. The meteorologists analyze these models, toss out anomalies, and generate a forecast. This whole process takes four hours.

The BVS programmers spend an additional five hours plugging the NHC forecast into their own system before issuing their report. As a result, BVS's forecasts are based on raw data recorded more than nine hours prior, often longer. The company that produces BVS doesn't want to advertise that fact. Buried on page 101 of the BVS-7 manual, it says: "The forecast model run and data preparation time is approximately 9 hours." That's the only tip-off to users that BVS's models weren't fresh.

For a quickly evolving storm like Joaquin, those hours were critical. The most serious consequence of the lag time was that earlier NHC forecasts, the source of Davidson's BVS reports, had positioned the storm as much as one hundred miles too far north, unlikely to intensify. Due to its unusual southwestwardly heading, the Atlantic's high temperature, and weak shear winds, Joaquin proved one of the most difficult storms to predict in recent history.

From September 30 to October 1, each NHC forecast incrementally shifted the hurricane's eye farther south and shifted its path from northwest to southwest. The changes were deliberately subtle because as a rule, the NHC prefers to avoid making major adjustments. Huge shifts would be unnecessarily confusing to the millions of people relying on these forecasts.

BVS duplicated that inaccuracy in each of its reports. On dawn of October 1, BVS was still showing the hurricane's eye eighty miles north of its actual location.

Davidson didn't know that the BVS information he relied on was always stale. He probably never received formal training on the system, and he may not have read the manual.

In between NHC reports, however, the Miami meteorologist on watch crafted a Marine Weather Discussion to explain his reasoning. Through the language of this discussion, people who live and die by weather—the world's pilots and sailors—could glean a much better sense of how reliable the forecasts are.

One of these interim reports was issued at 3:17 a.m. on September 30: "Joaquin is forecast to continue slowly west-southwest through the next 48 hours then turn north and accelerate," the forecaster wrote. "There is considerable uncertainty among major models with the details of track . . . intensity and timing not only of Joaquin but also the surrounding environment."

Between the morning of September 30 and the evening of

October 2, the NHC also provided a set of "key messages" via its Tropical Cyclone Discussions for Joaquin on its various social media accounts. These announcements detailed NHC's lack of confidence in its track and intensity forecasts for Joaquin and focused attention on the expected direct, and indirect, effects of the hurricane on the Bahamas and the eastern United States. South Carolina was certain to see major flooding.

Unfortunately, all of these more nuanced forecasting products were only available to those with internet access. The officers aboard *El Faro* had no way to browse the web.

A few mariners I spoke with said that they used the National Weather Service's (NWS) FTP site to download this information at sea. To do this, they had to email a coded request to the NWS, which would then automatically email them reports. It was an exceptionally clunky system.

We know from *El Faro*'s download records that Davidson didn't receive these supplemental reports. Davidson also may not have known that BVS offered up-to-date cyclone locations to its users, if they specifically request them by checking a box within the program. At the very least, we know that he didn't use this service on *El Faro*'s final voyage.

Instead, Davidson relied on weather information so outdated it was practically useless. Although the NHC shifted its forecasts over time, Davidson refused to shake his belief that Joaquin would turn north. The BVS reports only hardened his conviction.

Davidson thought he knew where the storm was and where it was going. Like a race car driver, he cornered tight along Joaquin's presumed course in order to shave off seconds from his time.

The other officers aboard weren't sure why the BVS and NHC forecasts were so different, and they didn't seem to be aware of the delay in the BVS reports—but at some point, it didn't matter. In situations like these, it made sense to respect the worst-case scenario for what it was: a possibility. Plan for the worst and hope for the best.

"That's where we should be and that's where the hurricane's gonna be," Danielle said, laughing, always laughing, at the absurdity of their predicament.

"Gotta be a better way to go," Jackie said.

"Max winds 85, gusts to 105 knots," she read out loud.

"We're gonna get our ass ripped."

"We are going to go right through the fucking eye."

"Kiss those containers good-bye."

CHAPTER 10

CAPTAIN MICHAEL DAVIDSON

Michael Davidson grew up on the coast of Maine where sea captains once ruled both land and sea. Their stately homes—built on majestic sites overlooking the water—still stand as reminders of a glorious shipping past. Driving up Route 1, you can't help noticing their prominent manses, many of which have been converted to inns named for the captains who built them: the Captain Swift Inn in Camden; Searsport's Captain A.V. Nickels Inn; and Kennebunkport's Captain Jefferds Inn. All summer long, No Vacancy signs hang from sturdy wood posts.

In 1829, eleven-year-old John P. Nichols, for one, left his Maine home for a life at sea. When he retired as a captain thirty-six year later, he built a 10,000-square-foot, twelve-bedroom Italianate home with multiple fireplaces, servants' quarters, and gilded accents, all topped with an airy cupola in Searsport overlooking Penobscot Bay. Nichols's house is now an inn, listed for sale in 2017 for $829,000, triple the price of most area homes.

Searsport was a natural place for the nineteenth century

captain to settle down. During Nichols's lifetime, the town was renowned for producing intrepid ship's officers and swift, multi-masted schooners. Now it's a few sharp bends in the road. You can pick up a Dunkin' Donuts coffee at the big gas station on the hill or buy some scratch tickets at Tozier's Family Market, which looks like it hasn't been touched since the Johnson administration. A modest maritime museum, closed in winter, hints at the town's illustrious past.

Arguably, the days of the respected ship captain ended with the age of sail. Now, mariners tell me, people on land consider them "ocean truckers"—working-class people earning police officer's wages, who move cheap underwear from China to your local Walmart. The mystique of life at sea has vanished.

On ships, however, the captain is still called "master."

As a boy, Davidson loved being on the water. He was a strong swimmer—a daredevil who impressed his friends by diving into the foam churned up by propellers of departing boats. His father, Sam Davidson, was the go-to accountant for Portland's fishermen, and a founding board member of the city's fish exchange. He was the proverbial big fish in Portland's small pond, which gave his son a sense of entitlement at an early age.

Mike earned his degree from Maine Maritime Academy in 1988, then took a job as a deckhand on the Casco Bay Lines ferries that shuttled people and cars around the small islands along the Maine coast. The ferries docked in South Portland within earshot of Mike's father's office. Eventually, Mike served as a deck officer aboard ConocoPhillips tankers in the Pacific Northwest. That job was his life, but he lost it over a dispute with a fellow officer. He then sailed with TOTE in 2005 for five years, took an interim job, then rejoined TOTE in 2013 as a third mate. When TOTE fired Captain Hearn as captain of *El Morro*, Davidson assumed his role. When that ship was scrapped in May 2014, Davidson became master of *El Faro*. By September 2015, due to the ten-week on/ten-week off schedule, he'd sailed on that particular vessel for less than one year.

Davidson made a good living, enough to maintain a forty-one-hundred-square-foot home on a cul-de-sac in a quiet Portland suburb for his wife, Theresa, and their two athletic daughters. But now that the girls were in college, expenses were high. He had two BMWs in the driveway, tuition bills, taxes. Mastering container ships paid up to $150,000, if you could get the work. Mariners who knew Davidson told me that he often complained that the money never went far enough. Like many middle-class Americans, Mike couldn't quite get his head above water.

Some people are born lucky, some not. Mike Davidson's mistakes had a bad habit of catching up with him. In his twenties, he got a string of speeding tickets. And one time, he grounded a Casco Bay's ferry on a shallow shoal. He had to ask a local fisherman where he was. If you knew the area, a Portland-based sailor told me, you'd never wander in there.

As a captain, Mike rubbed some of his colleagues the wrong way. He liked to have control over his ship and ran it according to his own highly regimented schedule. Davidson was a stickler for rules and safety protocols, but his meticulous nature, and need for control, could be misinterpreted as arrogance. One mariner told me, "I had friends who sailed with him. Things didn't always go so well. He had to move on. Bravado. That's all I can think of. You gotta check that at the gate when you join the ship. Do it at home. You can't do it on the ship. *I'm the captain.* Yeah, everybody knows you are."

Although he demanded respect from his officers, he had a reputation for spending most of his time in his stateroom; it was rumored that he constantly played video games. The best captains spent their time working closely with their officers and crew, running drills, and training those below them. Davidson's lack of engagement communicated apathy which, in turn, undermined his authority.

Managing a commercial ship requires serious skills—you're working with people with a broad range of education levels and economic backgrounds. Your officers may be distinguished maritime academy grads with families; some of your crew might be rough loners with high school degrees who only ship out to dry out. A major rift can develop between the engineers toiling below and the deck officers working above, as each views the other with suspicion and contempt. An even greater split can fall along racial lines. The demanding 24/7 work schedule, especially on a short liner run like the Jacksonville to Puerto Rico route, can exhaust the crew.

A weak leader can provoke defiance or complacency.

Davidson was exquisitely sensitive about how others saw him and just as defensive. In May 2015, two young coast guard officers took a few rides on *El Faro* and made observations about the crew as part of their training. One, Kimberly Beisner, testified that during one of those rides, she'd written in her workbook under "morale" that the crew seemed to simply do their jobs and return to their rooms. Davidson bristled at what he thought she was implying about his leadership skills. He told her that *her* presence made the crew uncomfortable. He said that usually they played cards and watched TV together, but since Kimberly came aboard, they were acting differently.

When his comments to Kimberly got around to the crew, many of them went out of their way to privately contradict the captain and reassure the young officer that no, *she* didn't make them uncomfortable; Davidson did. He was a hothead, got angry on a dime. If anyone was making the crew uncomfortable, it was the captain.

Life seemed to continually deal Davidson bad hands. He thought people were out to get him. He wasn't always wrong about that.

Shipping is brutal, especially at the top of the food chain. Officers get paid according to rank, not experience. Since the bottom fell out of the oil industry a few years ago, lucrative tanker jobs like Davidson's ConocoPhillips gig dried up. Pipelines are taking away more of those choice shipping routes.

When you reach the top of the heap—captain of a container ship—you can't relax. You watch your back because that chief mate standing next to you may be waiting for a chance to knock you off your block. A lot of mariners are forced to sail below their license—captains work as second mates, chief mates sail third—because they need the work, need the money, or can't stand the sound of guys sharpening their knives.

Throughout his career, Davidson was convinced people were trying to edge him out of promotions or undermine his authority. When he thought another officer at Conoco was pushing him out of his captain's job, Davidson left a message on the guy's voice mail threatening to kill him, which the mate simply played back to the company's human resources department. Davidson lost his job.

His worst fears came true again in 2015.

That spring, he had interviewed with TOTE for the position of captain aboard one of their two brand-new ships, and he considered himself a strong contender. The ships were a new class of vessels called Marlins, built in San Diego, designed to burn liquid natural gas (LNG) and diesel. They were commissioned by TOTE in 2012 to replace the heavy-oil-chugging *El Faro* in the Puerto Rico trade. They were just about ready for their sea trials, and TOTE hoped to get the first one fully operational by the end of the year.

TOTE's older steamships, like most of the world's deep-draft vessels, burned bunker fuel, the dregs of the petroleum industry. Similar to asphalt, bunker is nearly solid at room temperature and needs to be preheated to flow into a vessel's fuel lines. It has been targeted by the International Maritime Organization (IMO) as a significant source of greenhouse gases and harmful particulates.

The IMO, the United Nations' maritime regulatory organization, is the most influential body over this global industry, and the majority of countries adopt its laws and standards, with the notable exception of the United States, which cherry-picks recommendations. The group focuses its efforts on problems that plague the wild world of shipping: safety and environmental impact.

The maritime industry produces 2.2 percent of the anthropogenic carbon dioxide building up in our atmosphere, the gas that causes global warming. Sulfur dioxide, another oil by-product, has been linked to asthma and various lung issues. The IMO set a sulfur cap on commercial shipping vessels for 2020, essentially outlawing the use of bunker fuel, and the United States adopted even stricter environmental sulfur emissions standards.

In short, *El Faro* and her sister ships wouldn't meet these new requirements unless they were converted to diesel or LNG. The fate of *El Faro* was sealed. She would go into dry dock in the Bahamas in November 2015 for regular servicing and then sail to Alaska to temporarily relieve one of TOTE's more modern ships there. No doubt she'd get scrapped after that.

Clearly, being consigned to follow *El Faro* to Alaska was a career dead end.

Throughout the summer, Davidson waited and wondered whether he would master one of the Marlins. Other officers around him were getting plucked for training for the LNG fleet but no one knew who was next; staffing shifts seemed to come frequently, and at random. The chosen ones signed nondisclosure agreements with TOTE, promising to keep their promotions quiet, which only fueled suspicion and distrust aboard the working ships.

Meanwhile, Davidson's chief mate, Ray Thompson, wined and dined the crewing manager. She had joined TOTE ten years ago as a secretary for the company when it was based in New Jersey and stuck around long enough to be promoted into a position for which she may not have been properly trained.

When TOTE slashed upper management in the name of efficiency, the crewing manager stepped into a power vacuum created by corporate restructuring and was basking in the attention. Overzealous downsizing had created a critical managerial gap at TOTE: no one was officially in charge of evaluating the ships' officers. The port captain once did that work, but TOTE had eliminated the position years ago. TOTE's two port engineers were expected to issue regular assessments of officers, but due to general management confusion, they stopped doing it around 2013.

As a result, captains and mates were getting precious little feedback from the company on their performance. They didn't know where they stood.

That lack of assurance created unbearable tension on *El Faro* when TOTE began plucking officers for positions on its new ships.

The crewing manager took it upon herself to influence officer selection. She got her intel from the unlicensed crewmen who she regularly hired from the union hall. She'd already established a strong rapport: the Jacksonville-based unlicensed seamen adored her and so did their families. She was a white woman working alongside TOTE's powerful white male executives and her heart was with the black seamen. She looked out for them and socialized with them. Although she employed whomever the union hall sent for each shift, it was okay with her if they thought she had hiring power.

Word got around that she was a key decision maker, or at least had the ear of upper management, and would use any available information to advise her superiors on which captains and mates to promote. Once the unlicensed crew got wind of this, they had power over their commanding officers.

Damning information became currency aboard TOTE's ships. On *El Faro*, rumors up and down the command chain began churning. One AB caught an officer sleeping on watch. Instead of going through official channels to address the problem, he photographed his superior and spread the photo around. The situation was toxic. The quasi-military order necessary to the merchant marine had been breached.

The crewing manager seemed to have had a grudge against TOTE's ship officers. She was responsible for booking their flights down to Jacksonville and had a reputation for sending them on tortuous routes at odd hours with multiple connections. It was hard enough for mariners to leave family for months at a time. The crewing manager's apparently unsympathetic attitude to the officers' plight added insult to injury. When officers paid out of pocket for required training during their shore leave, they'd have to beg, sometimes for months, for the manager to reimburse them for their time. The officers' wives generally disliked her; the perception among them was that she was rude on the phone at best, vindictive at worst.

During the spring and summer of 2015, Florida-based chief mate Ray Thompson frequently visited the crewing manager during the day and took her out to lunch. He had the time to spare and was willing to do the legwork to get what he wanted. He had a family to support too.

While Davidson was under consideration for the new ships, cutting information about him began trickling out. In an email in May to TOTE's executives, the crewing manager reported that she had "dwindling confidence in his abilities as a leader overall." On July 8, TOTE's port engineer Jim Fisker-Andersen sent an email to TOTE's vice president of shipping operations, Phil Morrell, that said, "[Davidson's] a stateroom captain, I'm not sure he knows what a deck looks like, period. Least engaged of all four captains in the deck operation. He's great at sucking up to office staff."

Davidson may have hired a lawyer to pressure TOTE to give him a second look, and in August, the executives decided to promote Davidson. Just as they were preparing to give him the good news, another email suddenly appeared from the crewing manager describing an incident aboard *El Faro* that had occurred a month prior. It went something like this: While the ship was docked in Puerto Rico, one of the unlicensed crewmen showed up for work after partying in San Juan, clearly intoxicated. When he was confronted on the gangway, he went nuts, possibly wielding some kind of weapon. That's the kind of thing that happens in shipping—mariners can be a rough lot. Some have criminal records. Others have substance abuse problems. They get drug-tested, but that's not going to catch your garden variety alcoholic.

Davidson asked his chief mate, Ray Thompson, to check it out, so Thompson went down to the dock. Words were exchanged, no one got hurt, but the crewman was forbidden to board the ship in his condition. *El Faro* departed for Jacksonville without him. For some reason, Davidson didn't officially write up the incident, maybe because he wanted the whole thing to go away. The seaman needed to dry out. Maybe Davidson decided to cut him a break. Maybe he didn't want more trouble from the other crewmen. Maybe he didn't know company protocol for dealing with incidents like this.

It's possible that the disgruntled crewman himself complained to the crewing manager about his treatment by Davidson, changing the story to paint himself as a hapless victim. Regardless, this was exactly what the crewing manager needed to further undermine Davidson's candidacy. She leaked the story to Morrell in August, just as Davidson's promotion was going through.

A couple of weeks later, Davidson wrote a carefully worded email to TOTE's management requesting an explanation for why he'd been passed over and how he could improve his performance to optimize his chances for such a position in the future. No reply was documented.

Later, under oath, Morrell made a convoluted reference to the affair as a way of explaining why the captain suddenly lost favor with TOTE: "In my judgement going back to [TOTE CEO] Phil

Greene, [the incident] wasn't executed correctly, and [Davidson] really never spoke to [the crewman], and I recommended to Phil Greene that we—that there were just some administrative issues there. That I think we needed to stick with the course that was originally determined in which we work with some outside candidates for continuous improvement."

In early August, Davidson's relief, Captain Eric Axelsson, suddenly quit *El Faro*. TOTE had told the middle-aged shipmaster that the company was "going in a different direction" and didn't plan to hire him for the new ships. Axelsson may have also left because he was fed up with TOTE's lax attitude toward operations. He'd worked for years as a captain for Maersk and was accustomed to the Danish shipping company's precise, by-the-book management style. In contrast, TOTE's apparent indifference to its employees and operations standards annoyed the seasoned mariner. He walked off the ship and went home.

With Axelsson gone, TOTE asked Davidson to cut short his off-duty time and return to master *El Faro*. Clueless as to why he hadn't been promoted, Davidson seethed with anger, but took the job. Throughout August and September, however, he regularly pulled aside other officers and crew to complain about TOTE's decision not to promote him. Up to the end, Davidson was tormented by his fate.

Meanwhile, his own chief mate on *El Faro*, Ray Thompson, was promoted out from under him to master one of the new LNG ships. Davidson's worst fears had come true.

CHAPTER 11

QUESTION AUTHORITY?

26.07°N -76.16°W

Davidson was in a fragile state of mind, broken. He was disgusted that at age fifty-three, another job was slipping away.

Chief Mate Steve Shultz was worried about his future too, but at the moment felt compelled to focus on the fact that they were playing chicken with a major storm. The NHC reports had become unequivocal—Joaquin and *El Faro* were on a collision course. He had to say something.

In industries in which human error can lead to devastating consequences, it's important to foster good communication that respects the hierarchy while allowing room for debate. In the emergency room, nurses and doctors are trained to work together to make critical decisions about patient care. The commercial airline industry has developed standardized training procedures designed to encourage pilots and copilots to collaborate while in the cockpit. All military branches have checklists and protocols to help people work together to determine risk and check each other's work. No one makes decisions in a vacuum.

The US merchant marine is different. Few succeed in the shipping industry by questioning authority. Fresh out of school, cadets are no better than servants; they do what they're told. They might have to scrub and squeegee the cavernous holds of an oil tanker while choking on the fumes, or fetch someone's coffee while the ship rocks and rolls. A third mate never forgets that he's at the bottom of the heap, nothing more than a pair of eyes watching the sea and a pair of hands to assist the other officers. Second mates have the whiff of authority about them, but like the middle child, they're forever caught between giving commands and taking them. Even chief mates wait respectfully outside the captain's cabin until they're invited in.

On a ship, the captain reigns supreme. The final word. That's one thing that hasn't changed since Herman Melville wrote *Moby-Dick* in 1851.

"Dost thou then so much as dare to critically think of me?" cried Captain Ahab to his chief mate, Starbuck. "There is but one God that is Lord over the earth, and one captain that is lord over the *Pequod*."

Some masters solicit the opinions of their mates and actively encourage their officers to think for themselves. They're the teachers and mentors of the merchant marine committed to making their officers better mariners.

Captain Davidson had a dangerous swagger that belied a fragile ego. He jealously guarded his power and wrapped his authority around his heart like a shield.

Danielle knew that as second mate, she wouldn't be able to convince the captain to change course. Instead, she encouraged Chief Mate Shultz to give it a shot. To her, it was obvious that they were heading for trouble; armed with the facts, she assumed, he would have to act.

"Got the latest weather report from the SAT-C," she told Shultz when he returned to the bridge to take over watch after supper. "We and the storm are gonna be at the same place at the same time." She stood by waiting anxiously for the news to sink in so that they could turn the ship out of the path of the hurricane.

"Joaquin's tracking farther south and west than we saw in the last report," observed Shultz, examining the report.

"Yeah." Danielle nodded emphatically, ponytail bouncing. "It's right on our track line."

"Fascinating."

Not the response she was hoping for. Well, she'd done her part. She left the bridge to get some sleep before her midnight to 4:00 a.m. watch.

But Shultz had heard her. When Davidson arrived on the bridge after dinner, Shultz screwed up his courage. Clearing his throat, he said tentatively, "Did the second mate mention the weather?"

"Yeah," the captain grimly replied.

After some minutes of silence, Shultz said, "Get the BVS weather email yet?" He knew that Davidson didn't have much patience for the text-based NHC reports that came off the dot-matrix printer and was hoping the new BVS update would corroborate the other.

"I'll shoot it up as soon as I get it," Davidson said.

Ten minutes later, Shultz pushed further. It was already 5:40 p.m. Shultz had a feeling that the new forecast was sitting in the captain's in-box. "Is it . . . is it five o'clock that the BVS report comes in?"

"Nah, it's 18:00."

The chief mate thought for a moment. He knew he was right. The last report had come in around eleven o'clock that morning. Six hours later meant they should have had a new one at five o'clock. In fact, that particular BVS report was available for download at 5:03 p.m.

Davidson wouldn't admit he was wrong. But after being pressed further, he finally said, "I'll go down and check."

As chief mate, Shultz was in a tough spot. He had to step gingerly around his prickly captain.

He already had his hands full trying to figure out how to manage the crew.

Shultz had spent much of his career sailing in Alaska where the crews are often of Middle Eastern descent. He hadn't worked much with North Florida's black population, a group that, for generations, has had to deal with crushing racism.

According to some studies, white Florida is more racist than any other state—an attitude rooted in the state's late entry into slavery. Under the Spanish, Florida was a haven for fugitive slaves from Georgia and Alabama. But when the territory became part of the United States in 1822, opportunistic settlers rushed in to establish cotton and sugar plantations. By 1845, half of the state's population was enslaved by these parvenu slave owners, the dregs of the slavery establishment. Some say that the word *cracker* comes from the crude Florida slave owners' fondness for the whip.

Like many former slave regions, Florida had trouble adjusting to post–Civil War America. Racism festered in the Deep South and especially the Sunshine State—isolated by virtue of its geography, it bred a particularly virulent strain. Between 1900 and 1950, more blacks were lynched in Florida than any other state. The KKK thrived there. Black towns suffered raids and massacres perpetrated by whites. Jim Crow laws maintained strict segregation until Congress passed the Civil Rights Act in 1964, but many white Floridians still use the n-word without hesitation. As a result, members of its black communities have learned to stick together and distrust the white folks while outwardly deferring to them.

America's deep-seated racial divide has always played out on a microscale in the US Merchant Marine where close quarters aboard ships mean close encounters with prejudice. Far from shore, simmering anger can erupt into violence. Everyone has to be vigilant.

On *El Faro*, tensions between white officers and mostly black crew increased exponentially after the 2012 shooting of seventeen-year-old Trayvon Martin in Sanford, Florida, and the subsequent exoneration of his Latino killer, George Zimmerman. The tension further escalated when, in 2013, TOTE ordered its officers to clamp down on overtime. Everyone's paycheck was affected, but as the manager of the unlicensed crew, the chief mate stood on the front lines of this wildly unpopular shift in compensation. The mostly black ABs viewed the change as another indignity forced on them by their white bosses and became increasingly passive aggressive.

The ship's boatswain's job was to mediate between the chief mate and the crew. If the boatswain wasn't plugged in or committed, things could get difficult aboard the ship. Work would take longer. Rumors would spread. Fuses got shorter.

By 2015, hostility and resentment aboard *El Faro* were at an all-time high, presaging the shift in national political rhetoric from President Obama's conciliatory tone to President Trump's white supremacist leanings. (Florida would go on to vote for Trump.)

The white administrators in TOTE's Jacksonville office were well aware of the racial issues aboard their ships. In a memo regarding Steve Shultz's candidacy for chief mate sent on July 30, 2015, the crewing manager wrote that she planned to brief him on the "divide and conquer with regards to the crew cooperation, etc."

Shultz was about to be thrown into a hopelessly complex situation.

Divide and conquer? Where would he start?

Shultz didn't want to compromise his authority. But he also didn't want to overtly question his crew's competence. He simply wanted to trust that the people he was hired to manage did their jobs well, while he took the time to learn about the ship. He was working hard to establish order and camaraderie among his fellow officers, too. Especially Captain Davidson.

Alone on the bridge, Shultz's attention turned to his helmsman. Frank Hamm looked sullen and worried. Shultz tapped the big man's shoulder. "You all right?"

"Me?" Frank said, shaking off his bad thoughts.

"Smellin' the weather?"

"Yeah," Frank said. "Looks like it's buildin' up."

"Yup," Shultz said. "I'm gonna log it as Force Six." (On the Beaufort scale, that meant 22- to 27-knot winds, up to thirteen-foot seas.)

Frank was used to watching everything that happened on the bridge and keeping his mouth shut. That's how he kept his job. But when he looked at the BVS projection, he got nervous enough to say something. "It looks like that doggone picture you showed me this mornin' when we was on the outskirts of that yellow," he said. "If that's the eye, we're right here, getting close."

Meanwhile, the SAT-C printer started grinding away again with the latest news from Miami. Shultz ripped off the paper and studied it.

"What'd it say?" Frank asked anxiously.

"They're saying a bunch of numbers, positions, times, millibars," said Shultz condescendingly.

In fact, what Shultz read scared him. Joaquin had rapidly expanded and intensified since he'd last looked at the weather—now it had swelled into a big morass feasting on the unusually warm waters of the Atlantic.

"It looks like we gettin' close to that eye," Frank said, hoping that the officer standing in front of him would challenge the captain.

"I'll make a recommendation," Shultz said.

Frank chuckled. Making a recommendation sounded so namby-pamby. What they really had to do was turn the ship around. Get the hell out of there.

But Shultz's job was to stand by his commanding officer and

not foment insubordination or panic, so he downplayed the situation they were in. "I've rode these ships through horrible storms," he told Frank. "I'm not scared to go through it. We're stable. Remember in *The Perfect Storm* with that container ship rocking and rolling and rolling?"

"I saw that but I don't like watching that," Frank said, laughing uncomfortably at the thought of those monstrous CGI waves. "I saw it but when I seen it, I did not like that movie. 'Cause I'm on that ship, you know?"

Shultz kept needling his nervous AB: "I got pictures of these ships with all them trailers out there layin' on their sides—hangin' over the railings. I mean literally trailers hangin' over the sides with chains hangin' down in the water."

"I ain't never seen these containers with chains dropping, destroyed. Know what I'm sayin'? I ain't been in anything, nothin' like that before. Knock on wood I won't be a part of nothin' like that." Frank had no need to wrestle with a hurricane to prove he was a man. Life was hard enough. He was there for the paycheck and the peace at sea, not to test his grit.

"I've seen water chest-deep down there on the second deck," Shultz continued, now thinking about the ship riding under him. "And that's why I said those scuttles need to be dogged, not just flipped down. You know—they need to be spun and sealed."

On *El Faro*'s wide-open second deck, you could wander the narrow alleys between the parked trailers and containers, stepping over a spaghetti of thick electric cables that kept the refrigerated ones chilled in the Caribbean heat. Close to the hull on both sides of the ship were three pairs of manholelike hatches, called scuttles. A mate could climb through them and descend a ladder to take a look inside the vast holds to check lashings to make sure nothing had come loose.

If a ship is sealed up with all her hatches closed tight, she can survive nearly any storm, thanks to the fact that air is lighter than water. Like a plastic bottle with its lid screwed on, you can push it under and it'll always pop back up. You could even load it with rocks, and as long as the weight-to-air ratio remains below a certain point, it will float. Add enough water though, and that ratio shifts. Add enough water and the thing will sink like a brick.

In other words, if you've got buoyancy—if you're not overloaded—you can get slammed by anything nature throws at you, and you'll still float. Your cargo, though, might suffer great indignities.

El Faro's mammoth cargo compartments were dimly lit and crowded—a vast lot of trailers chained like wild animals to the deck crouching in eerie silence. But there was plenty of air in those holds, each acting like an enormous steel balloon.

Huge hydraulic watertight doors running the width of the ship segmented the ship's length into compartments so that flooding or fire could be contained. At sea, all five two-story holds on *El Faro* acted like individual air pockets vital to the ship's buoyancy. But the man-size doors cut into the hold doors were sometimes left open after the ship was loaded. If a cargo compartment flooded and these small doors weren't closed, water would get everywhere.

Cut into the second deck, the six scuttles were equipped with heavy steel lids, similar to manhole covers, hinged along one side. To make them watertight, you'd have to spin a wheel on top to engage the dogs—thick metal prongs that keyed into the sides—like a bank vault door. Some of *El Faro*'s hinged hatch covers required up to fifteen turns of the wheel to secure the dogs, but this number varied. It could be half that, depending on the scuttle. A lot of times, the deckhands just flipped over the lid and gave the wheel a few spins. But that didn't guarantee they were locked.

Could a hatch fly open? In 1929, British writer Richard Hughes went to sea as an able seaman and nearly died when his ship's wooden hatches blew apart in a storm. "When a hurricane blows off the roof of a house," he later wrote, "it does not as a rule get inside the house and burst it from within. The flow of the wind over the roof makes a vacuum on the lee side, and so sucks it off. When the *Archimedes* heeled over away from the weather, her deck made an angle very similar to the lee side of a roof: therefore the suction this wind exerted on it must have been terrific. But decks, of course, are enormously strong. Hatches, on the other hand, are not."

As the hurricane brutalized Hughes's ship, the high winds turned air into water: "Though no heavy seas were coming on board, the spray was so nearly solid water that hundreds of tons would find their way below in a very short time."

Before *El Faro* hit bad weather, a vigilant mate would have made sure those scuttles were closed tight. Water was bound to wash over the deck, so batten down the hatches, and triple-check them. But crew used these scuttles a lot; it wasn't easy to track when the doors were left open during a voyage. And some of the scuttles may have been too banged up to seal shut anyway.

The ship had lots of large openings in the hull on the second deck, many of which had been blocked with steel plates to prevent pirates from boarding when *El Faro* served in the military. Shultz worried that they'd given the crew a false sense of security. "Yeah, they won't keep the water out," Shultz said, thinking about *El Faro*'s blocking plates. "They might be good for breaking waves but they're not watertight. They're just like breakwaters. They're all bent up, beat up. It's not like you can lock them or anything. Get a lot of green water over the bow, this thing is gonna crack in half. It's scary."

Shultz knew their best defense was challenging the captain's route.

He got his chance when Davidson came up to the bridge a little before 7:00 to share the two-hour-old BVS report. It had caught up somewhat with the other forecasts, but still underestimated Joaquin's intensity.

"BVS says the same thing as the NHC report," Shultz said, observing the brighter crimsons and purples of the hurricane's dangerous center spreading deeper into the region. It showed the hurricane continuing its southwest course, trapping them against the shallows of the Bahamas. They were running out of open water. If Joaquin kept coming for them, they'd have nowhere to go but rock. "Got it going farther down south than the last time." Davidson didn't respond.

Silently, Shultz formulated a way to suggest a course change that wouldn't offend his master.

Clearing his throat, he finally said, "Um, would, would you consider going the other side of San Salvador?"

San Salvador was 124 miles south of *El Faro*'s current position. It's thought to be the first land Christopher Columbus encountered when he sailed to the New World. An underwater monument marks the spot where the *Pinta* is said to have anchored. Traveling at 18 knots, they'd reach its lee in six hours, around one o'clock in the morning. Easy sailing down there for a deep-draft vessel. The sixty-three-square-mile windswept scrap of sand is actually the wispy remains of the peak of a fifteen-thousand-foot-high submerged mountain, wasted away by storms and seas to little more than a sliver of dry land. All around it, the ocean floor bottoms out two and half miles down.

"Yeah, I thought about that," Davidson said distantly, gazing out at the deepening gray.

But inside, he seethed. When he'd gone down to his stateroom to download the latest BVS report, he discovered an email from TOTE sitting in his in-box. The director of labor relations was asking whether Davidson wanted to take a week off when he got to San Juan. The offer sent a chill down his spine. Shipping companies didn't interrupt a captain's tour of duty for no reason. Pulling a captain was disruptive to the pace of the ship and the crew. What was TOTE trying to tell him? The whole thing seemed suspicious.

TOTE was churning through its captains; no one's job was safe. Axelsson had unceremoniously resigned from the company in August after being told that his services were "no longer required." Standing next to Davidson was a chief mate he barely knew, one of many cycling through these days.

The turnover was so disruptive that Jeff Mathias, the extra chief engineer aboard *El Faro*, had written an email to Tim Neeson, the port engineer, in August: "We're on our third new chief mate this month and he's still trying to find the mess deck."

Davidson himself was considered a temporary hire, although he'd been working for the company for a couple of years.

"I don't know what the deal is here," he grumbled to Shultz, obsessing over the hidden meaning of the email. "It said, 'Captain so-and-so needs to get back to work.'"

Was TOTE going to put a new master in his place and never call him back? Davidson had written back to Neeson declining the offer—he'd serve until his time was up on December 30. But he couldn't help being bedeviled by the unsettling email.

Shultz stood by awaiting an answer about their route. Should they duck behind San Salvador? Davidson knew that wisp of sand wouldn't cast much of a lee, but it would get them slightly farther west of Joaquin. It would be tricky sailing around there, though. They'd be threading the needle in the deep but narrow trench between San Salvador and the Bahamas Bank in high winds and high seas. The heading change wouldn't take *El Faro* much off course, so what the hell, Davidson thought, and agreed to the plan. Might as well make a minor detour if it made Shultz happy.

The chief mate began entering the new waypoints.

Davidson watched Shultz work, heavily brooding over his

future. "When TOTE lays this ship up, they're not gonna take us back," he told Shultz.

"I hear what you're saying, Captain," Shultz said. "I'm in line for the chopping block."

"Yeah. Same here," said Davidson.

"I'm waiting to get screwed."

"Same here."

"I don't know what's gonna happen to me," Shultz added. A chilly wind came from the captain.

When Danielle came up to relieve Shultz for dinner, Davidson told her about their new route to San Salvador. He assured his young second mate that he'd be there with her on the bridge through it all.

"That's what I was aiming for—lots and lots of deep water," Shultz triumphantly chimed in, relieved that the captain had acquiesced. "We can straight-line it outta here. Keep this distance off it."

Davidson did have another option that wasn't discussed. He could have hove to right there—turned 180 degrees and followed *El Yunque*'s path back toward Key West. There was no guarantee that Joaquin wouldn't chase them all the way back to Jacksonville, but if Davidson turned around he had the option of sailing west if he had to, through the Straits of Florida. Beyond that, he had all of the Gulf of Mexico at his disposal. *El Faro* was fast, much faster than the storm.

But Michael Davidson wanted to prove that he was TOTE's kind of man. He didn't back down. He dug in. He'd sail that ship right into San Juan on Friday morning, just as planned. "See, there's all sorts of outs," he proudly told his chief mate, taking charge. "You don't necessarily have to go down the Old Bahama Channel. There's lots of places to duck into."

Plotting out their new course, Davidson confidently concluded that they'd get to Puerto Rico before eight o'clock in the morning on Friday. He went to his cabin and sent an email to TOTE that said as much.

Davidson would prove to TOTE that he was a bold leader, the kind of captain who could outsmart a hurricane and deliver the goods on time. *El Faro* could take it. They'd get there, even if they had a rocky ride. He'd show them all that he was a new class of man for a new class of ships.

CHAPTER 12

THE JONES ACT

The two great oceans that lap at America's shores have shaped our collective consciousness in myriad ways. *Moby-Dick* may always be the greatest American novel. Paintings and models of clipper ships, three-masted schooners, and frigates in full sail still adorn the walls of the country's law firms, boardrooms, and government buildings. Common American English is rife with nautical words and phrases—*skyscraper, close quarters, give a wide berth, high and dry, know the ropes, slush fund,* and *fly-by-night.*

Many people can trace their roots back to ancestors who arrived in the New World by boats, powered by sail or steam.

Sailing ships symbolize economic power born of equal parts folly and freedom—folly in that we willingly leaped into the great unknown; freedom in that with ships, we could travel anywhere in the world or get dashed on the rocks trying.

Shipping is in our blood and baked into our laws.

America's earliest European settlers sailed here from England and clung to the coast where they eked out a living from the rocky soil while awaiting sight of sails. They did what they could to praise God and carve out a devout life from a lawless land, willingly accepting their dependence on a steady influx of ships to maintain a modicum of civility on the edge of a great wilderness.

When New England's soil revealed itself less than suitable for farming, eyes turned to the sea. Massachusetts Bay colonists built a few boats and tentatively dipped them into the North Atlantic. The ocean might have spit them out or drowned them, but it held promise for the hardiest men and women. With enough practice, intrepid mariners discovered plentiful fisheries rife with cod. Some outfitted their boats for longer voyages, spawning a robust shipbuilding industry that attracted French, British, and Dutch buyers.

Successive generations embraced the economic freedom that fishing and shipping offered, and they proved quite adept at both. Soon, young America's merchant vessels were ubiquitous in every ocean.

It was shipping that first pried open the great divide between north and south, a cultural difference once best understood a thousand miles offshore. New Englanders had little to sell but an intrepid spirit, so they became global merchants, hustling goods across many oceans, creating intricate financial instruments to spread profit and risk. Meanwhile, Europeans' insatiable demand for cotton and tobacco kept southern planters anchored to the land; they generally left complex trans-Atlantic logistics to others. As long as the ships—most often British—came with slaves and departed with crops, southerners had no need to complain, or rebel.

New Englanders were global citizens. Southern planters waited for the world to come to them.

America's first merchants didn't just transport goods from here to there; they were inveterate speculators. New England shippers bought their own cargo at the lowest possible prices and sold it off wherever its value might be highest. Aboard every vessel was a supercargo—a trained broker who earned a share of the voyage's profits. His job was to haggle and plot on behalf of the ship's owners. He would chart the ship's course for the best payoffs—buying fish and pine from New England's fishermen and farmers; demanding hefty returns in the Caribbean islands in the form of molasses and mahogany; then heading to European ports hoping for big payouts for these highly coveted raw goods.

Colonial merchants also opened up new markets to trade. Their unquenchable thirst for rare and pricey goods sent them dodging their European competitors around the treacherous Cape Horn to California and the Pacific Northwest, where Native Americans supplied them with luxurious furs, and on to Asia, especially China, where they traded these furs for even rarer spices and silks.

The shipowners, often small consortia of men from a single New England town, may have assumed all the liability of these risky voyages, but also reaped impressive rewards, if and when the vessels returned. Shippers crewed their boats with men and boys who, through profit-sharing, enjoyed an economic mobility unknown in Europe. A twelve-year-old Maine boy could leave the family farm to cook in a ship's galley and work his way up to captain by the age of twenty-five, then retire onshore at age forty wealthy enough to run his own fleet. By the mid-1700s, fine silver, porcelain, and furnishings began flowing back to the New World.

Throughout the eighteenth century, a robust shipbuilding industry blossomed, attracting the best engineering minds to American shores, so much so that master builders in London petitioned their government to write laws that would discourage talented shipwrights from emigrating. Regardless, they went, drawn to the booming economy abroad where virgin forests provided tall, strong wood for masts, cladding, and hulls, and hemp created strong, pliable rope for rigging and sails.

When the Nantucketers perfected the whaling ship, installing try-pots directly onto its deck so they could refine the blubber at sea, they were able to sail around the world, sometimes several times over, filling their stores with fine oils that fetched high prices, which, in turn, financed more whalers and whaling men, as well as handsome homes and new industries ashore.

From the first, America's seamen were rebels. They constantly defied British law, sailing into any ports they pleased, trading with whomever they liked. Such violations were generally tolerated until George III's ascension to the throne in 1760. During his rule, the English Parliament became increasingly frustrated with the colonists' lack of respect, and revenue. Displeased with their zeal for free trade, clearly in violation of the Crown's restrictive Acts of Trade and Navigation, the British clumsily clamped down, issuing harsher tariffs and penalties. In response, America's great shipping families, merchants, and lawyers pooled their resources to battle the king across the pond.

The American Revolution was as much a maritime dispute as it was a philosophical battle. The wealthiest colonists—the captains and merchants—would not be legislated and taxed without representation.

By 1776, America's vast fleet of merchant and fishing boats, expertly handled by some of the world's best seamen, as well as their world-class shipbuilding knowledge, was poised to play a critical role in the Revolutionary War. Though the merchant vessels generally fell short as battleships, wrote the popular historian Samuel Eliot Morison in 1921, they proved excellent at foiling the British navy's efforts in other ways. They were a constant nuisance—intercepting communications, forcing the enemy into rocky shores, seizing vessels and cargo. The king quickly learned that fighting a war across the Atlantic against such savvy foes was an exercise in frustration.

In the early days of the United States, shipping tariffs made up 90 percent of the federal revenue. Accordingly, the Founding

Fathers had an intense interest in protecting the new country's shipping families. "Maritime interests were supreme," wrote Morison. "The Constitution of 1780 was a lawyers' and merchants' constitution, directed toward something like quarterdeck efficiency in government and the protection of property against democratic pirates."

When drafting the United State's earliest laws, Alexander Hamilton relied on a brain trust of shipowners—the so-called Essex Junto—a group of influential New England merchants who championed neutrality with Britain after the war to protect their assets on the high seas. Hamilton supported their financial interests; in return, they buoyed his efforts to establish the federal treasury and Bank of the United States. From that alliance emerged America's first cabotage law, "An Act for Registering and Clearing Vessels, Regulating the Coasting Trade, and for Other Purposes," in 1789. It ensured that the domestic coastal trade was exclusively open to American-owned ships, a sweet deal for America's shippers. Many countries now have such laws as well.

Hamilton had other reasons for supporting the country's commercial fleet any way he could. In his mind, the idea of a merchant marine—a fleet of privately owned ships crewed by American seamen ready to deploy in defense of country—presented a compelling public-private partnership opportunity.

That notion would prove critical during the War of 1812. The turn of the nineteenth century found England at war with Napoleon. American merchants, always looking for an edge, embraced the role of war profiteers—brazenly trading with the Crown's enemy, demonstrating little loyalty to the country that begat them. What's worse, British seamen abandoned their naval posts to join the lucrative American ships, undermining His Majesty's sea power. When the British navy began seizing American vessels and impressing sailors in retaliation, President James Madison asked Congress to declare war.

The young country was not prepared. Big-government foes, especially Thomas Jefferson, had lowered taxes, reduced the military to a few soldiers, and allowed Hamilton's national bank charter to expire, leaving precious few resources for the government to fight England. Madison struggled to rally support and floundered.

This would be the first test of America's merchant marine as defender of American interests. Comprising one of the largest fleets in the world, the country's commercial vessels departed ports as privateers—bounty hunters sanctioned by the US government—with the rallying cry, "Free trade and sailors' rights!" Financed by speculators ashore, they seized foreign vessels, captured the sailors aboard, and sold the pirated goods for tremendous profit. Privateering was a business venture, not a military mission, but it wreaked havoc on Britain's ability to fight the Americans. (Even so, the British managed to burn down the White House in 1814.)

As shipping evolved from sail to steam, the federal government further defined and regulated its maritime industry, but it wasn't until the creation of the Merchant Marine Act of 1920 (commonly known as the Jones Act) that these laws became fully codified.

The Jones Act shaped the modern American sailor and the modern American shipper in profound ways. Rooted in the country's earliest laws, the Jones Act is unabashedly protectionist. It shelters domestic shipping from the pressures of the global market by upholding an America-first monopoly. To participate in the US trade along its coasts and interior waterways, ships must be owned and crewed by US citizens, and they must be built in domestic shipyards. Today, nearly all the country's goods and petroleum products spend some time aboard US-registered boats, manned by a twenty-seven-hundred-odd-member seafaring corps, the great majority of them union men and women.

The Jones Act also offered American seamen unprecedented rights. Whereas the British navy ran by "Rum, sodomy, and the lash," quipped Winston Churchill, American seamen were suddenly permitted to sue shipping companies for negligence (if mariners could prove that their employer put them at undue risk) or unseaworthiness (if the seaman could prove that the substandard condition of the vessel caused injury). This was a triumph of maritime's labor leaders who had fought more than three decades for protections for the men who served on merchant ships.

As the twentieth century marched on, the union's hard-earned grip on the industry put it at odds with conservatives and anticommunists. William Randolph Hearst planted his tabloids with antiunion and antimariner messaging. Ronald Reagan, who called the merchant mariners "little children," worked throughout his tenure to reduce their public benefits.

The Jones Act has inherent economic costs: According to the World Economic Forum, banning foreign vessels from carrying cargo between US ports costs Americans at least $200 million each year. Some studies estimate that Americans pay $.15 more for every gallon of gas, thanks to lack of shipping competition. Just before being diagnosed with a brain tumor in the summer of 2017, Senator John McCain introduced legislation for the umpteenth time to repeal the Jones Act, calling it "an archaic and burdensome law that hinders free trade, stifles the economy, and ultimately harms consumers. The protectionist mentality embodied by the Jones Act directly contradicts the lessons we have learned about the benefits of a free and open market."

Defenders of the Jones Act view it as economically and strategically crucial to the American way of life. The Jones Act bolsters the domestic maritime industry by guaranteeing buyers for American-made ships and American-trained sailors. These days, that's significant: Ships built stateside can cost twice as much as those manufactured abroad. The creators of the Jones Act recognized these discrepancies and created generous federal tax subsidies to soften the blow.

Ultimately, it's a small price to pay for a standing quasi-navy, ready to deploy at a moment's notice, which made all the difference during World War II when the US government conscripted America's merchant fleet and seamen to transport troops and munitions to Europe under risk of U-boat fire. This was exceptionally dangerous work; the US Merchant Marine suffered more casualties than any other military division in the war.

But heavily regulating and subsidizing an industry can have unintended consequences. Until midcentury, the US Merchant Marine dominated the seas. Now the dwindling number of American-flagged ships comprise merely 4 percent of the world's fleet. The shift is due to simple economics: it's cheaper to register ships in countries that have looser regulations, like Liberia and Panama.

Plenty of American-owned ships fly foreign flags to get around Jones Act requirements. This flight among shippers from America's registers, called "flags of convenience," is an invitation to anarchy, argues William Langewiesche in *The Outlaw Sea*. By registering a ship elsewhere, shipowners can avoid US taxes, labor laws, domestic construction requirements, and anything else that cuts into profitability. And oftentimes ships are registered to landlocked nations, which underscores the free-for-all nature of this system and the Wild West nature of the sea. When a non–Jones Act ship goes down, Langewiesche notes, there's little risk to the customers and shipping companies. The hulls and cargoes are fully insured and actual ownership is obfuscated by a rat's nest of LLCs and shell corporations.

With the passing of the Jones Act in the 1920s, Puerto Rico had the potential to become one of the few lucrative runs in the shipping industry: a hungry customer completely dependent on imports via Jones Act vessels. But first, the island needed to develop a robust middle class—a new generation of Puerto Rican consumers who were wealthy enough to buy American goods. In 1950,

Congress enacted Operation Bootstrap, a stimulation package that would check several boxes. The tax-based incentives gave mainland manufacturers a new source of cheap American labor—Puerto Ricans; it created a new market for American stuff; and all this would result in a booming new trade route for US-flagged ships. Bolstered by these local and federal tax credits, mainland corporations raced to offshore manufacturing jobs to the island. Billions of dollars flowed to the island from the United States to build plants where lower wages meant things could get built cheaper, without the penalty of import duties.

From the island's sugar cane economy emerged an industrial one, complete with a thriving middle class reliant on imported food, clothing, and oil. *El Faro* was built for that first Puerto Rican boom.

New federal tax incentives introduced in the 1990s brought a second wave of investment to the island, this time in high-tech manufacturing. At its height, Puerto Rico was one of the top pharmaceutical producers in the world.

When those tax incentives expired in 2002, the companies pulled out. With them went jobs, and then the people—Puerto Rico's population plummeted from 4.5 million in 2002 to 3.6 million by the end of the decade as islanders in search of work moved to the US mainland.

In spite of economic collapse, the island couldn't wean itself off imports. In a tropical paradise where some of the world's best coffee grows in the mountains, many of the people still enjoy their highly processed, packaged, and imported Nescafé. Despite being awash in sunshine, Puerto Rico depends on imported petroleum products to run its electric plants. The island's Electric Power Authority reportedly pays as much as 30 percent more than the United States for LNG due to restrictions on foreign-flagged ships, yet efforts to convert to solar power have been stymied by a system that has strong allegiances to the oil industry. Puerto Rico was quickly buried in a mountain of debt.

Vulture capitalists vacuumed up the island's massive debt under usurious terms, a solid bet considering the fact that Puerto Rico is prohibited by American law from declaring bankruptcy. This little known technicality became law when a tiny "with the exception of Puerto Rico and Washington, DC" amendment got tacked onto a bankruptcy restructuring bill passed by Congress in 1984. Among other things, the bill specifically exempted the island from bankruptcy protections all other American cities enjoy, so its debts must be paid, even if it means people starve and schools close. The origin of this exception is obscure, though some claim that South Carolina senator Strom Thurmond slipped in this seemingly minor point right before the bill became law.

Exploiting the ensuing chaos, Colombian drug cartels shifted their trafficking route into the United States from Mexico to the struggling island. By 2010, DEA agents were finding drugs on the ships from San Juan—in containers, on the crew, and packed in cars in the cargo holds—while gang-related crime on the island skyrocketed.

Strapped for cash, the Puerto Rican government, a major employer of islanders, implemented austerity. But the one thing Puerto Rico couldn't save money on: ultra-high Jones Act shipping costs. Like the early American colonists, Puerto Ricans had found themselves at the mercy of the ships.

In 2017, that dependency would prove disastrous when Hurricane Maria hit the island, wiping out roads, power lines, and houses. Relief supply deliveries were hampered by Jones Act regulations, severely limiting the number of ships permitted to service the beleaguered island. Overnight, the law became national news as media outlets reported on the island's frustratingly slow recovery. The opinion pages were inundated with calls for a repeal of the Jones Act, so much so that President Trump put a temporary ten-day waive on the law to allow international freighters the opportunity to deliver necessary goods. But by October of that year, public outrage dissipated, leaving the Jones Act's grip on Puerto Rico's economy intact. As of December 2017, the *New York Times* estimated that at least 1,052 Puerto Ricans had died as a direct result of the island's handicapped efforts to reconstruct infrastructure.

CHAPTER 13

EVENING

The sun dropped like a stone on Wednesday evening, September 30, casting *El Faro* in darkness. At eight o'clock, Third Mate Jeremie Riehm joined his AB Jack Jackson on the bridge for their four-hour watch. Chief Mate Steve Shultz went over the newest course change—a welcome concession from Captain Davidson. They'd maintain a heading of 150 degrees—south-southeast straight toward the lee side of tiny San Salvador.

Captain Davidson stood nearby, listening closely. "Safety of the ship comes first," he reminded them, enunciating each syllable. "San Salvador's gonna afford us a lot of lee," Davidson told them. He continued referring to the hurricane—now a Category 3 with 85-knot winds—as "the low."

Jeremie had been assiduously watching the forecasts on the TV in his room. "I just hope Joaquin's not worse than what BVS is saying," he told Davidson, "because Weather Underground is saying it's a lot. They're saying it's more like 85 miles per hour, not 50 like on BVS. They're saying this is much more powerful than BVS is saying right now."

"We'll be passing clear on the backside of it," announced Davidson, dismissing his third mate's intransigence. "Just keep steaming. Our speed is tremendous right now. The faster we're going, the better. This will put the wind on the stern a little bit more. It's giving us a push."

Jeremie considered San Salvador, a six-by-twelve-mile island, just a wisp of sand, highest elevation 140 feet. It stood all alone in deep ocean. He knew it wouldn't block Joaquin's wind, though the seas would be somewhat mitigated as they broke on the island's eastern shore. Then he thought about how *El Faro*'s aging power plant might fare in the storm.

Deep in the bowels of the ship, the forty-year-old engine burned ton after ton of bunker fuel, producing astounding heat (900°F) that converted water to superheated dry steam. The pressure coming from that steam, about 860 pounds per square inch, spun the engine's first turbine at a rate of 6,700 revolutions per minute. That's way too fast for a ship's propeller to spin (more like a jet engine) so a series of gears stepped down this rate of rotation to a comfortable 132 RPMs. At the end of a massive horizontal shaft spun the ship's 25-foot-tall propeller. This whole system was bathed in thousands of gallons of lube oil to prevent the fast-spinning metal parts from breaking down under friction and splintering into a million shards.

All that machinery occupied a three-story engine room built on the floor of the hull. Unlike today's clean diesel engines, it was dark, gloomy, and dirty down there. Narrow steel ladders and catwalks led workers around countless steam pipes, valves, and pumps, many of which were unmarked, or poorly lit. Earlier in 2015, while the ship was under way, an AB accidentally closed the wrong valve and shut down the engine. The chief engineer quickly identified the problem and got it up and running in about ten minutes. But that experience shows how easy it is to disrupt the system.

Anyone who's opened the hood of a car would recognize most parts on a ship's diesel engine—cylinders, pistons, crank shafts. But a ship's steam engine is a strange and awesome beast. It's easy to get disoriented in the three-story assemblage that sits heavily on the floor of the hull. To operate a steam engine, you have to know it cold.

Taking care of the old engine required constant vigilance. *El Faro*'s chief engineers' turnover notes to each other were often several pages long, detailing everything they fixed, everything they had their eye on, and everything that needed immediate attention.

Rich Pusatere was the thirty-four-year-old chief engineer aboard *El Faro* on September 30. He'd last boarded the ship on August 11, leaving his wife and baby daughter at home in North Carolina. Rich had trained at SUNY Maritime in Fort Schuyler and, like Jeff Mathias, had a passion for steam. The newer diesel engines weren't nearly as exciting. They didn't demand the constant attention he lavished on his beloved turbines and boilers.

Rich's parents had done everything they could to make sure he had opportunities they hadn't. They were children of Italian immigrants, and they'd escaped New York City tenements for a tiny suburban house in a tiny village north of the city along the Hudson River. Rich's father, Frank, had an engineering mind and ran plants for ConEd. Then he ran a chimney sweep company. Now he works as a forensic fire investigator. Rich took after his dad—when his mother, Lilian, totaled her car, he rebuilt it in their garage at the age of sixteen.

As Rich approached his high school graduation, Frank started to hound him about going to college. Rich didn't have a clue what he wanted to do. Well, said Frank, I think you should go somewhere. He did some research for his son and discovered the SUNY Maritime program in Fort Schuyler—a tiny spit of land under the

Throgs Neck Bridge in the Bronx. Rich could easily commute from their Croton-on-Hudson home, and the school was affordable.

Rich was intense. He drove himself hard and expected everyone around him to do the same. He was only interested in logic and facts. "Officers talk to people. Engineers talk to machines," seamen have told me, and that was never truer than with Rich. He was happiest when he was covered in grease fixing something in the sweltering heat of the engine room, earplugs in, earmuffs on.

When his daughter, Josephine, was born in 2013, Rich's awkwardness with the infant made his parents laugh out loud. He wasn't exactly a cuddly kind of guy.

Chief engineers can be otherworldly. They knock a valve here or bang a pump there, and things just click back into place. The old salts who cut their teeth on steam know every single component of the system so well that they can draw it from memory. They scramble up and down the steep metal ladders and catwalks that thread through the three-story behemoths to get to their lube pumps, sumps, and manifolds. When something goes wrong, they can usually pinpoint the problem in less than five minutes. But knowledge like this takes time to acquire. You have to pay attention, and really care, because no one's going to teach you everything you need to know.

Maritime schools can't fully cover the intricacies of operating a steam engine. It's just not worth the time, especially because there are so few of these engines still being used. Instead, learning is done on the job. If the chief is impatient, the kind of person who likes fixing things himself without having to explain every little thing, it will take longer for his knowledge to get passed down.

"It's a dying art," says Brian Young, a former mariner, now an investigator with the National Transportation Safety Board. "There aren't a lot of steamships anymore. Take an old chief who knows what he's doing and it's magic. He just comes down, tweaks something, and walks away and everything's fine. I'd gone over to the Persian Gulf in Desert Storm and they broke out this old chief who couldn't even leave his office because he was so old and had a limp. He smoked cigars all day long. And if he had to come down, if something was really bad, he would limp down there and he would swish through the smoke, sit down, and turn some valve halfway. Halfway. And everything would be fixed."

Rich worked tirelessly to keep *El Faro* running, and sometimes got frustrated when the people around him didn't push themselves as hard. On his 2015 evaluation of first engineer Keith Griffin, a man he had to work with closely, Rich couldn't mask his disappointment. He felt that Keith didn't have the same drive and passion.

That evaluation drove a deep rift between the two men.

Rich may have felt the same frustration with Davidson.

Sailing into heavy seas tests the most critical relationship aboard a steam ship: that between captain and chief engineer. Though separated by a dozen flights of stairs, the pair represent the mind and heart of the ship. They must collaborate to get her through—you can't will a ship through gale force winds. If a captain decides to steer any vessel into the belly of a hurricane, the engineers must be prepared to assiduously monitor her health to gauge how much power she can give. As the external forces on the ship grow, all the captain's senses should be attuned to the vessel's reactions to wind, wave, and rudder, while the chief engineer obsessively monitors steam pressure, oil pressure, and RPMs. At the moment she falters, shows signs of distress, that bridge phone should ring.

But weatherwise, it had been a quiet couple of years in the Caribbean, and so the captain and young engineers aboard *El Faro* hadn't been seriously tested. In the engine room, the recent maritime school grads had spent their time taking orders from their chiefs—fixing, patching, and oiling the giant machine—executing their duties, learning what they could from their supervisors.

A powerful storm would expose every weakness on the vessel.

Steam, the wondrous force that drives the turbines, travels through a complex web of steel pipes throughout the engine room under intense pressure. These pipes are secured by hangers. If one fails, shaken loose by the rocking and rolling of a ship, the pipe could rupture, exploding into a thick fog of scalding steam. We know that *El Faro*'s steam pipes hangers were in bad shape. The list of things to be done when she went into dry dock in November included: "Verify all pipe hangers, snubbers, and supports for the main steam system in the engine room are in good order to take the increased pounding in Alaska service. Many hangers are currently loose or broken."

The condition of *El Faro*'s boiler itself looms large in this story, too. A month before her final voyage an inspector wrote: "Burner throats have deteriorated severely especially between number one and three burners. Cracking and loss of material plus heavy buildup of fuel is present on all three throats. The front wall of the starboard boiler is in very bad shape. It is highly recommended that the front wall tubes as well as the brick including burner throats be renewed on both boilers. The brick is moving due to soot buildup behind them as a result can cause casing fires as well as damage to tubes to the point of failure."

The inspector hoped that the boiler repairs would be done right away, but these fixes were deferred for the November dry dock.

Those were problems people could see. On a 790-foot ship, there was a lot that inspectors couldn't see.

All this was on Jeremie's mind as he arrived to take over the watch at 8:00 p.m. That, and securing life rafts. Before he came up to the bridge, he'd put a couple of car straps on one of them to ensure it didn't fly off the second deck. He also secured the small first aid room.

In contrast, Davidson's biggest concern seemed to be their arrival time. "I told the office to expect us at eight o'clock in the morning in San Juan," he told Jeremie. Then, turning to amiable

Jack, Davidson said, "Before it got dark, we altered our course. Picked a new route to get farther away from the low," still refusing to call Joaquin a hurricane. "This course will put the wind on our stern which gives us a nice push. So fasten your belt till we get on that backside of it."

He reassured Jeremie that he would be there for him: "I will definitely be up for the better part of your watch. So if you see anything you don't like, don't hesitate to change course and give me a shout."

Before leaving the bridge, Davidson said: "I have the BVS new waypoint route sheet. I'm gonna enter it in on my computer. I'll bring it back up." Then he disappeared in his stateroom and failed to return to the bridge for eight hours. The waypoint route sheet never came.

In the darkness, Jeremie and Jack were left alone to face the consequences of Davidson's decisions. Jeremie again wondered aloud about the discrepancy between the reports he'd seen. The Weather Channel and the NHC forecasts aligned; BVS was the outlier.

"This BVS map tells us that in the morning, we'll see 53-knot winds," he said. "But if the Weather Channel is right and BVS is wrong, then we're gonna be really close to 90-plus-knot winds. It's more powerful than we thought. We can't outrun Joaquin, you know? The hurricane is supposed to hook north right here, but what if it doesn't?"

"What if we get close?" Jack asked. "We get jammed into those islands there, and it starts comin' at us."

"That's what I'm thinking," said Jeremie. He roughly combed his fingers through his hair like a backhoe clearing out a ditch. "Maybe I'm just being a Chicken Little."

Jack struggled to understand how they'd found themselves in this mess. He decided to blame it on the breakdown of TOTE's organization. Some really good, seasoned men like Haley and Villacampa had been fired. Others got away with murder. TOTE was a mess. Jeremie wasn't paranoid.

He reminded Jeremie that one chief mate on *El Yunque* had been repeatedly caught sleeping during his watch on the bridge and no action had been taken. "Then he got caught *again* and nothing happened. Asshole kept on doing it. Got away with it and nobody noticed, nobody cared. Nobody else notices anything and then I'm always the bearer of bad tidings. I'm known as the troublemaker and that's how I've learned to keep my mouth shut at those safety meetings. They don't wanna hear anything you got to say, so don't say anything."

The mariners he sailed with were getting dangerously complacent, following orders instead of asking questions. Jack thought back on the time he sailed on a car carrier and wanted to know which vents he should close if they needed to dump CO_2 on a fire to snuff it out. "The vents weren't even labeled. So we decided to close all of them. And we went out there to test them and they were rusted solid."

"Yeah," Jeremie said. "You're gonna have leadership that's conscientious, and you're gonna have guys that aren't. I call them out and sometimes it gets me in trouble. Our primary responsibility out here should be making sure it's a safe ship. It's hard out here because there's so many fucking things to address. It's like, where do you start, you know?"

Jack returned to the situation at hand. He wanted to be prepared: "If you gotta divert if that hurricane veers south quickly, how much time do you have to set up waypoints, if you gotta duck inside?"

"Well, right now," Jeremie said, "we got nowhere to go. We can head south, you know. But it's a good idea to have an alternate. We should have a backup route."

"Yeah," said Jack. "There's no escape plan."

"We need a plan B. If this happens, we do this. If this happens, we do this. Because already twice now, we've changed course to avoid Joaquin."

Jack worried about lack of experience aboard the ship: "Remember when I came on back in February and was asking about the hurricane history? And everybody's just shrugging—*Nah, whatever*—you know? *We never see anything. Ever.*"

"They've got short memories," agreed Jeremie. "Some other captain would have taken one look at that and said, *We're gonna go the Old Bahama Channel. We're not taking any chances here.* And we'll go well south of it, and we'll be getting in a bit late. We'll be off schedule but we'll catch up."

"That's what I thought we were gonna do," said Jack. "Now we're boxed in—just islands on one side and a hurricane on the other side. You got yourself a collision here."

Jeremie caught himself again. He didn't want to second-guess Davidson in front of a crew member. "This guy's been through a lot worse than this. He's been sailing for a long, long time. He did it up in Alaska."

Jack wasn't so sure. "I'll never have faith in these assholes like I used to," he said. "One captain sailed us right into a storm. All naive and confident. Oh sure," he said sarcastically, "*they* know what they're doing. As I got thrown to the deck, I could hear the captain screaming. We were going over this big wave. First the ship started pitching and rolling. Then it got worse and worse. Then all of the sudden it was like some kind of wild animal trying to break out—like a bull in a stall. Going wilder and wilder and wilder and pretty soon it was out of control. Then all the cargo broke loose and oh my God.

"We were rolling, and a wave came in and slammed us. We came to this shuddering stop. I mean, I was sure we were going over. Positive.

"It was like death, actually. It was like we were fated to die. No one spoke about it for two to three days. People were shell-shocked. Everybody felt death was right on us. It was like this presence. Bring your rosary beads."

"Well, you didn't say anything then, and you're here to tell the tale," said Jeremie. "Guess I'm just turning into a Chicken Little, but I have a feeling like something bad is gonna happen."

At 11:00 p.m., he ripped the latest NHC forecast off the SAT-C printer and couldn't keep quiet anymore. The third mate picked up the phone and called Davidson. "Hey, Captain, sorry to wake you. The latest weather just came in and I thought you might want to take a look at it, if you have a chance. Just looking at the forecast and looking at our tracking, which way it's going, thought you might wanna take a look at it."

What do you see? Davidson asked him.

"Well, it's the, the, the current forecast has max winds at 100 miles an hour at the storm's center. And it's moving at 230 [degrees] at five knots. I assume it'll stay on that same direction for, say, the next five hours. So it's advancing toward our bow, and our course puts us real close to it. I could be more specific. I could plot that out. But it's gonna be real close. We're looking to meet it at, say, four o'clock in the morning."

The captain asked him to draw it on the chart. Jeremie did and six minutes later, called him back: "At four o'clock this morning, we'll be twenty-two miles from the center of the hurricane with max [sustained winds] 100 [miles per hour], with gusts to 120 miles per hour and strengthening. From what I can see, at 2:00 a.m., we could head south, and that would open up some distance between us and the hurricane." Then Jeremie caught himself, remembering to defer to his commanding officer. "I mean, of course I'd want you to verify what I'm seeing."

He told the captain in no uncertain terms that if they continued on their current course, they would sail right into the most dangerous area of the hurricane—the eyewall, and into the eye. "Just so you know," he told Davidson. "That's how close we'll be."

Jeremie listened to Davidson's response and hung up the phone. He turned slowly back to Jack, shaking his head. "Well, he seems to think that we'll be south of it by then. So the winds won't be an issue."

"Fuck," Jack said.

"Fuck."

"We'll be twenty-two miles from the center."

"Fuck."

"He's saying we'll be in the southwest quadrant of the hurricane, which is the safest place to be if you're going to get close. Wind will be coming from the north and it'll push us away, toward Puerto Rico. I trust what he's saying, it's just I don't like being twenty-two miles away from 100-knot winds. This doesn't even sound right."

"No matter which way it's hittin' ya, it's still 100-knot winds," Jack agreed.

The captain may have been unhinged for steering them into this storm, but then again, hubris was part of his job. Great captains were made by taking risks and living to boast about them. Shipping history is replete with slightly mad masters aching to test their mettle, pushing their vessels and crew to the brink and coming out heroes. In the modern age, everything always turned out fine. *El Faro* had been inspected, tested, and secured. And they had a wide range of safety equipment at their disposal.

After he got off the phone with Jeremie, Davidson did not check the weather forecast. He did not download the latest BVS forecast waiting for him in his in-box. Instead, he went to bed. Records show that Captain Davidson did not download the eleven o'clock BVS forecast until five o'clock the next morning. His third mate had told him the truth. But he didn't want to hear it.

CHAPTER 14

NIGHT

El Faro was small and alone in a vast, black emptiness, shrouded by Joaquin's miles-thick clouds. The few lights on the ship's console were like cold, distant stars; a laptop-size radar screen regularly burst into silent fireworks representing the agitated surface of the sea. Another screen tracked the ship's position, their intended course shown as a simple straight line. The instruments would reveal other vessels in the area, too, if there were any.

"We don't have any options here," Jeremie said to Jack. "We got nowhere to go."

"Jesus, man. Don't tell me anymore. I don't even wanna hear it," Jack told him.

The two men stood behind the console squinting into the abyss. Invisible waves pounding the vessel's port side bullied her into a long, slow, rhythmic side-to-side roll, not enough to make them sick, but enough that as they walked around the bridge, they ran their fingertips lightly along the grab bars to stay upright. Small things began to shake loose and slide across desktops onto the floor. Anyone below who was awake flicked on cabin lights and stowed away pens, coffee cups, and eyeglasses in drawers and cabinets.

The distant hum of the engine far below was the only source of comfort, like a heartbeat.

Just before midnight, Danielle entered the bridge in a flood of light.

"Good morning, Danielle," Jeremie said.

"Good morning," she said, closing the bridge door softly behind her, sealing the room once again in the heavy gloom. She approached the console. Jeremie winced in the darkness. He wasn't sure he wanted to leave the helm to his fellow officer; so many things could go wrong. But he wasn't sure he wanted to stay, either. After experiencing hours of anxiety, he was mentally exhausted.

"I'll show you what's going on back here in the chart room," he said, pushing aside the blackout curtains that protected their night vision. He slipped into the warm glow of the dim chart room light. They stood together, hunched over the table, the large map of the area laid out in front of them.

Jeremie picked up the wax pencil resting on the Plexiglass protecting the paper chart and used it as a pointer as he spoke. "This is our latest course," he told Danielle.

They would travel thirty-two miles to the next waypoint—a predetermined coordinate that marked their course—which put them right on track to meet the storm, if the NHC's forecast was correct. "The NHC weather report that came in at 2300 was different from BVS," Jeremie noted. "I told that to the captain, but this is the course he wants."

"This is totally crazy," Danielle said, once more breaking into excited, uncontrollable laughter. Her one consolation was that in her lifetime, no American container ship had gone down. Ships sometimes got caught in vicious storms, but they always made it out. They'd be okay. Still, this was scary.

"If the forecast holds," Jeremie continued, "at four o'clock in the morning, you'll be twenty-five miles from the center and still going toward it." Twenty-five miles from the hurricane's eye positioned them square in the eye wall, the deadliest part, where the powerful forces driving the system are most concentrated. If he knew that, he didn't say it. Neither did Danielle.

Just then, AB Jackie Jones came up to the bridge to take over Jack Jackson's watch. "Hey, where's the party at?"

"It's gonna be a party in a few hours," Jack said lugubriously.

"I know," Jackie said. "I just seen a little TV, and then the TV went out. We fixin' to be right in the middle of the storm."

Jack suggested that his fellow AB check his survival gear.

"Do you know where your EPIRB is?" he asked Jackie, referring to the emergency beacon that all seamen were supposed to have on them so that they could be easily located by rescue crew.

"I ain't got no EPIRB," Jackie answered. "Uh-oh. This is a bad deal."

"When I get to my cabin, I'm getting out my flashlight, life saver, and Gumby suit," Jack said, using the slang term for an immersion suit—a thick neoprene unibody outfit required for every mariner aboard a merchant ship sailing in winter in the North Atlantic. The suit comes with built-in booties, mittens, and a hood designed to stave off hypothermia in cold water. When you're zipped up and Velcroed inside, you look like a red version of its 1950s cartoon namesake, minus the Play-Doh eyes.

"I put all the Cokes and water bottles in the cabinet too," he told Jackie. "Probably gonna start throwin' shit around here. We're gonna drink the water if they aren't careful. We're gonna get sucked into it."

"We're the only idiots out here," Jack remarked before heading down to his cabin.

Meanwhile, at the chart table, Danielle and Jeremie tried to reconcile Davidson's orders with the latest NHC forecast. "This is the second time we changed our route and the storm just keeps coming for us," Jeremie said. "But the ship's solid."

"Yeah," Danielle said. "I trust her."

The two pored over their charts, searching for a way out of the hurricane's path if it didn't stop coming for them. One option was to continue south toward Crooked Island Passage. It was a straight shot, once they got past San Salvador, via the West Channel to the Old Bahama Channel—nice, wide, and deep. At that point, they could keep running west, away from the storm, or hook around Cuba and head west or south. Two small reefs stood in the middle of Crooked Island Passage, so whoever was at the helm then would have to maneuver carefully around hazards, steaming in high seas through unknown waters in the dark, relying heavily on their paper charts and depth finders. If they had electronic charts, it would have been easier to visualize their path as they worked their way through.

Danielle knew Davidson. He wouldn't readily embrace a plan that risked grounding them on shallow shoals.

"What's concerning me is that the information we're getting from other sources is so much different from BVS," Jeremie said again. "According to everyone else, Joaquin wants to intensify and keep going southwest."

"Blowin' now like hell," Jackie called out from the helm. "It's just sittin' out here suckin' up the heat outta these waters."

As an officer, Danielle had to stand by her captain's decisions when talking to her AB. She certainly didn't want to cause widespread panic. Turning away from Jeremie, she reassured Jackie that they'd be fine: "We would've been within twenty-five miles of it, now we're just gonna dodge down to port, just southwest of the center." In her heart, she must have known this wasn't likely.

"Why don't we slow down? This game is stupid," Jackie told the night.

Jeremie headed down to his room to get some rest, leaving

Jackie and Danielle alone on the bridge. The second mate may have felt some relief that Davidson wasn't up there yet. She privately dreaded that midnight watch, and not just because they'd be wrestling with Joaquin. She didn't want to be alone with the captain on the bridge. She'd told her friends that she loathed him; her hatred must have intensified now that he was putting them in danger. She suspected he was the kind of guy who, when rejected, swung from hurt to fury in a heartbeat.

Danielle returned to the chart room to painstakingly plot an alternate course based on her discussion with Jeremie. She was riveted to her work, spending more than half an hour with her pencil and clear plastic triangle working out waypoints to avoid the reefs along their way, to get them safely down to the Old Bahama Channel.

At 12:41, she told Jackie, "I may have a solution. At two o'clock we'll be in a good spot to alter course south to 186 degrees. That course line, if we stay on it, keeps us five miles away from any kind of shallow area. Not a lot of wiggle room, but right now where we're going, we don't have much wiggle room."

"Then we'd have the wind about dead on our ass," Jackie said. A push in the right direction sure would be nice, but running with the waves requires excellent steerage and speed control. They'd need to hang behind the crest to avoid getting swamped as the wave breaks, a nearly impossible maneuver if they couldn't see the ocean.

"This course could get us away from Joaquin," said Danielle. "Go south, get away from it, and connect with the Old Bahama Channel, right into San Juan. Unless this damn storm goes further south. It keeps trying to follow us."

"It ain't gonna do nothing but sit down here growing, getting stronger and stronger," Jackie said. He liked her plan.

Outside, the sea growled in warning. Joaquin was ahead, bearing down on them, churning up the ocean and leaving behind a frothy wake. Yet its slow westward momentum had compressed its forward edge, meaning they wouldn't feel the full brunt of the system until they were much closer to its eye.

"It doesn't really feel like we're getting near a hurricane," Danielle said.

Still, they were approaching a monster, and the calm was disquieting. "It's an eerie feeling, I'll tell you," said Jackie, going over to study the latest forecast emerging from the NAVTEX printer. "They say this storm could grow to a Category 4."

By one o'clock in the morning, Davidson still hadn't come up to the bridge, nor had he called to check in with his second mate like he'd promised. Maybe he thought they were clear of any danger. "I don't know if he can sleep knowing all of this," Danielle said. Still, she waited.

The radar told them they were passing Cat Island on their right, and San Salvador was coming up on their left. Danielle had worked out a route that would save them. But she still hesitated to disturb the captain.

Instead, she turned on the satellite radio and heard a brief report from the Weather Channel: "Hurricane Joaquin's been upgraded to a Category 3 storm. It's expected to pass near the Bahamas before heading toward the East Coast of the United States. The powerful storm may bring ten inches of rain to some areas."

"Oh my God," she said. "Now it's a Category 3." A huge wave slapped *El Faro*'s port side.

"Whoa," said Jackie. "Biggest one since I've been up here. You can't pound your way through them waves. It'll break the ship in half."

At 1:20 a.m., Danielle finally picked up the ship's phone and dialed Davidson. It took him a few rings to answer. "I just wanted to tell you that I've charted another course that runs south of the island chain into the Old Bahama Channel," she told him. "Otherwise, we'll be meeting the storm. Fox News just said it's up to a

Category 3. It isn't looking good right now. My track line: at two o'clock, alter course straight south and then we'll go through all these shallow areas, and the next course change will be through the Bahamas and then just turn to San Juan."

There was a pause. They were in fourteen-foot seas by then, with 55-knot winds.

"Okay," she answered softly. "I'll adjust course to 1-1-6. Okay, thank you."

Davidson didn't just reject Danielle's plan, he ordered that she alter the ship's heading to 116 degrees, pointing *El Faro* even more easterly, a direct route to the eye. *El Faro* would soon find herself rammed broadside by forty-foot-high waves as 100-mile-per-hour sustained winds shoved the ship over to starboard. It was the worst possible decision.

"He said to run it."

In less than an hour, they began feeling the fury of Joaquin. Seas rose up around them as torrents of rain pummeled the windows like God's own pressure washer. Through the deluge, they saw flashes—either lightning off the bow or the reefers on deck shorting out. They could hear clanking but didn't know where it was coming from. As they rolled in the growing swells, their propeller continually spun out, causing their speed to drop down to 16 knots. Small things began flying free below them, as huge waves crashed over the decks.

The whipping winds made a deafening roar on the bridge, and Danielle had to raise her voice to be heard over the din.

Then cargo started breaking loose. They couldn't see it, but they could hear thumps in the dark. Danielle and Jackie had to lean to stand upright as the floor rhythmically pitched left, then right beneath them.

Danielle wanted to believe that Davidson was correct, that they would pass below Joaquin any minute. She was waiting for the winds to shift to their starboard side as they got south of the hurricane. When the screeching gusts continued pounding *El Faro* from the northeast, she fought back waves of fear.

Both weather reports—NHC and BVS—were wrong. NHC showed Joaquin's eye too far north by a few dozen miles, and BVS by double that. Davidson had taken BVS's forecast as gospel and sealed their fate. They'd cleaved so close to the hurricane that they'd left no margin for error. They were firmly in Joaquin's grip.

"I figured the captain would be up here," Jackie finally said at 2:47.

"I thought so too," Danielle said, struggling to hide her terror. "I'm surprised—he said he was gonna come up."

"He'll play hero tomorrow."

CHAPTER 15

NECESITAMOS LA MERCANCÍA

Shipping is the engine that drives the modern world. It's our T-shirts, diapers, beer, and crisp new sheets. It's everything we wear, everything we touch, the parts inside our cars, our laptops, and our dinner. More than 90 percent of the things people need come by ship.

Cheap transport has shaped the needs of every man, woman, and child on Earth.

It binds us together.

Its collapse could rip us apart.

Take away the mundane things, life's basics—the stuff we don't even think about—and civilization ceases to exist. A 2017 story in *The New Yorker* by Evan Osnos illustrated our dependence on the cheap movement of goods across multiple oceans. Osnos's piece was about super-rich doomsday preppers, tech wizards who spend a lot of time thinking about the end of our world. "Some of the wealthiest people in America—in Silicon Valley, New York, and beyond—are getting ready for the crackup of civilization," he wrote.

In Osnos's story, the thirty-year-old CEO of Reddit, Steve Huffman, said that he was "increasingly concerned about basic American political stability and the risk of large-scale unrest. 'Some sort of institutional collapse, then you just lose *shipping*—that sort of stuff.' (Prepper blogs call such a scenario WROL, 'without rule of law.')" Huffman believes that contemporary life "rests on a fragile consensus. 'I think, to some degree, we all collectively take it on faith that our country works, that our currency is valuable, the peaceful transfer of power—that all of these things that we hold dear work because we believe they work.'"

Huffman's point is that global shipping depends on an extensive, yet fragile, global infrastructure: cheap energy, satellite communications, international agreements, regulations, and policing. Doomsday preppers understand that these institutions are at risk. And while you might dismiss their paranoia, the more you know, the more you realize that they could have a point.

Americans have always harbored paranoid tendencies that continue to thrive. Some Americans are convinced that their own federal government poses the biggest threat to their way of life. Along with the taxman, the US Coast Guard can find itself in the middle of this argument. When its inspectors deploy to militia-rich areas like the Great Lakes, Alaska, and parts of the South, they sometimes have to go armed. They've come to inspect boats to ensure that the vessels comply with federal regulations—that they've met certain safety standards so that they're less likely to kill their owners or riders—but the antigovernment types consider that kind of maritime regulation a fundamental threat to their freedoms.

Like Huffman, Captain Michael Davidson was a self-proclaimed survivalist, a doomsday prepper, armed and ready to protect his kin. He had an arsenal of weapons in his house, plus a reserve supply of food and water, and maxi pads to soak up blood.

He was preparing for an apocalyptic event but probably didn't have just one scenario in mind. He trained for something catastrophic, and much of that training involved shooting his way out. Or learning how to survive in the woods after a fill-in-the-blank enemy burns down the local Target. He made sure his wife and daughters knew how to use guns and equipped them with Glocks. One time, he showed an *El Faro* officer a video of himself dressed in camo, armed with a range of automatic weapons. He charged a target guns ablaze, and when he ran out of bullets, he drew handguns, and when those ran out, he pulled knives.

Preppers anticipate the rise of independent militias hell-bent on taking down the government, and with it, the global economy. In the face of armed insurrection, they're betting that the government will fold, leaving every man for himself. Various congressional attempts at instituting gun control, especially following inconceivable acts of random violence, stoke those fires, banding preppers with antigovernment types, creating a single, united, paranoid movement. Best to be prepared.

The main problem with being a doomsday prepper is that when you're fully armed and ready for battle, it's very unsatisfying when nothing happens. If the threats aren't real, what good are all those guns? They secretly yearn for an opportunity to test their mettle. Otherwise, they've spent an awful lot of time, money, and energy preparing for nothing.

Some preppers invite standoffs with the law, hoping to achieve glory in a blitz of bullets. Militia members in Oregon in 2016, for example, occupied a national wildlife refuge for several weeks, but finally abandoned their post after running out of food and trashing the place. Their leader ultimately stormed a federal roadblock in his truck. To some, his was a martyr's death.

Traditionally, Maine-based survivalists have been quieter than their western counterparts, which may be because their contra roots run deeper. Much of Maine wasn't founded by patriots but

Tories—British loyalists who fled north when the American colonists went rogue and defied the Crown. As they have since the Revolution, they blend into their suburban or rural surroundings and hunker down, secretly nurturing their paranoia with weekend training shooting sessions, learning to distinguish poisonous berries from edible ones, or building a bunker out of birch saplings.

In America, the prepper movement is rapidly expanding. From Arizona to Michigan to Maine, survivalism has become a serious hobby. And in recent years, the demographics of the movement have shifted from rural crazy to wealthy, sophisticated, and urban techies whose doomsday scenarios—some involving nuclear war with rogue states like North Korea, for example—have become frighteningly real.

Captain Jason Neubauer of the US Coast Guard was stationed in Hawaii in 2011 when he learned firsthand what happens when the ships stop coming— the threat of anarchy. In March of that year, the archipelago was pounded by tsunami waves, the aftershocks of a deadly earthquake off the coast of Japan that killed nearly sixteen thousand people and caused Fukushima nuclear power plant to melt down.

The waves traveled 600 miles per hour across the Pacific, giving the coast guard time to shut down Hawaii's commercial ports. Within twenty-four hours, islanders began feeling the pain of a world without shipping. "No one realizes that Hawaii is running day to day," Jason told me. "We were getting calls from other islands saying they were running short on fuel." Pretty soon, they would feel the pinch of more critical shortages, like food and water. How long would it take before they had anarchy on their hands?

Like Hawaii, Puerto Rico is an island that depends on ships. Stuffed into containers often originating in Asia, clothes, electronics, and appliances are loaded onto the world's biggest vessels, transported to Los Angeles, unloaded and reloaded onto smaller ships bound for Florida via the Panama Canal. They travel up the coast to Jacksonville where they're transferred once again, along with bananas, chicken, cheese, and Doritos, onto ships or barges heading to San Juan. From there, goods may travel on to other American islands, like St. Croix, where after all that handling, half a gallon of milk costs $7.

The fate of *El Faro*—an elderly ship transporting increasingly greater cargo to Puerto Rico—shows how America's maritime laws, bent to accommodate corporate interests, can lead to tragic consequences.

Since she'd been built ships were getting bigger and bigger. The world's population had doubled. There was increasing demand for goods. *El Faro* had to keep up to be profitable. While her ability to take on trailers was useful, she needed to behave more like a modern container ship. That meant taking containers loaded from above by the port cranes.

It seemed like an easy change to make. In 2002, Sea Star (now TOTE) made a big bet on Puerto Rico. Saltchuk, the Seattle-based transportation and logistics company that owned Sea Star, had gone to great lengths to profit off its trio of elderly US-flagged steamships. It kept its workhorses running an average of twenty-eight years longer than any others docked in America's ports, sending them to the Persian Gulf during the two wars there, and into the rough Pacific Northwest trade back and forth to Alaska, another Jones Act monopoly. And now it wanted *El Faro* to carry more cargo to Puerto Rico.

In 2003, the company wanted to remove *El Faro*'s spar deck—an ancillary deck above the main deck designed to accommodate trailers—and shut off the ramp that led up there. Now trailers and cars could only drive onto the second deck and follow two ramps down to the decks below, instead of going up.

This change made a lot of sense from a commerce perspective. The world had fully embraced containerization since *El Faro* was built thirty years before. These days, all the container ships crossing the Pacific and the Atlantic are fat and wide to accommodate containers stacked as much as ten high.

But how would *El Faro* react to this dramatic transformation? Instead of carrying all the weight in her belly, she was being asked to shoulder a huge backpack, one that would get pushed around by the wind on stormy days. *El Faro* was designed to be a sprinter, sleek and streamlined. Her engineers gave her a sharp bow to slice a narrow path through the seas at 22 knots with the least amount of drag. Her elegant taper from deck to keel was very different from the bulbous modern container ships, which travel around 15 knots, meaning she was naturally tippier, and slower to right.

Could *El Faro* be both a beast of burden and fast as hell?

Obviously, containers piled high on her main deck shifted her center of gravity upward. The added weight also pushed all her openings—vents and hatches—closer to the water.

El Faro became trickier to model as well. The most stable ship is a loaded tanker. The oil in its holds acts as a low, uniform weight throughout the vessel, so it's easy to predict how that tanker will react under various sea and wind conditions. It's simple to model, thankfully, because if a tanker sinks, the environmental consequences are far-reaching.

El Faro's cargo presented a complex modeling situation. She would be carrying containers, trailers, cars, and liquid-filled tanks in a dynamic environment, the open sea. It's a nightmare of a physics problem. How that ship would react to changing conditions depended on everything—the shape of her hull, where the weight sat relative to her center of gravity, how much ballast was in her tanks, and the additional sail area (vertical surface), which would cause her to heel over when the wind pushed against it. Calculating all this requires a computer. You *could* do it by hand, but it would take hours and a small error could produce major miscalculations.

So TOTE's officers and port engineer relied on CargoMax,

standard ship-loading software customized for the ship's unique variables. Before *El Faro* was loaded for each voyage, the port engineer would get a manifest and start plugging in numbers. Each container's weight was provided by TOTE's customers, which he'd enter into CargoMax. Each trailer had a weight. Each car had a weight. After entering everything in, CargoMax would spit out a loading plan. Box A had to go here. Box B had to go there.

Loading had the aura of an exact science. But it wasn't. Possibility for error lurked everywhere. Was the customer properly reporting box weights? You had to trust them. And what was the center of gravity of those boxes? No one knew. Did the port engineer enter the weights correctly? It was a lot of work to get it right. Minor numbers got screwed up all the time, sometimes causing major problems. And on her final voyage, a wrong decimal threw off the calculations so much that during loading the ship experienced a serious list. Cargo had to be shifted around to correct it, delaying her departure.

And, of course, the computer model itself was just that: a model. It was designed with margins of error. But it wasn't designed to project what would happen if one of those containers broke loose in a hurricane. That was precisely the kind of dynamic situation that no one can precisely model. You'll just watch in awe as those containers shake loose, break free, and fall over the side, ripping the deck apart while pulling the ship over with their momentum.

The US Coast Guard declared that TOTE's proposed changes to *El Faro* were major modifications, and the vessel would have to be brought up to current safety standards. That ruling meant TOTE would have to swap out *El Faro*'s open lifeboats for the modern, submarine-like enclosed boats required on every ship built after 1986. Open lifeboats had been banned internationally because the small crafts had proven so vulnerable in heavy seas. Much of America's aging merchant fleet, however, did not follow all IMO laws, and still carried open lifeboats because the shipping industry lobbied against the regulations, which it considered draconian and egregiously antibusiness. (For example, it took decades of lobbying by Samuel Plimsoll to convince shipping companies to mark load lines on their hulls, indicating how much weight they could take before they were officially overloaded. If they could, one coast guard officer told me, shipping companies would grind those marks right off again.)

For three years, TOTE's executives and lawyers fought the coast guard for the right to keep *El Faro*'s fiberglass open lifeboats on board, and they eventually convinced the agency to reverse its ruling, arguing that similar work had been done on sister vessels without triggering a major modification ruling. In its defense, TOTE said, it would cost the company an additional $7 to $9 million or so to do all the required work. "If this is true," TOTE's letter to the coast guard said, "and we believe it is, the project may not get funded and that would be injurious to our company and the maritime community in general."

The US Coast Guard reversed its decision in 2006. And so, *El Faro* sailed with her outdated lifesaving equipment, along with her ancient boiler and questionable stability. The coast guard had delegated the stability analysis to the American Bureau of Shipping (ABS). There was no review of the calculations done by ABS's people that would have determined how the ship would pitch, roll, and react to various sea conditions under her new configuration.

A few more things slipped through the cracks: At some point, *El Faro*'s load line was raised two feet, allowing the vessel to take much more weight than ever before. And six eighteen-thousand-gallon tanks for transporting fructose syrup, totaling approximately 13,400 pounds, were permanently installed in 2014. All these changes permanently altered the ship's stability, but the Marine Safety Center—the coast guard's office dedicated to ship design and engineering—found no record of these modifications.

Was *El Faro* more vulnerable to sinking? In a word, yes.

Desperate to squeeze every dime out of its ships in a client-driven market, TOTE's upper management also broke federal law. Beginning around 2003, executives from the three companies in the Puerto Rican trade—TOTE, Horizon, and Crowley—clandestinely met to divide up the route's major customers and set shipping rates, agreeing not to compete with or undercut one another. This price-fixing continued for at least five years until a whistleblower alerted the federal government in 2008.

Following a five-year series of suits, the US Justice Department fined the companies a total of $46.2 million for violation of the Sherman Antitrust Act and sentenced six midlevel shipping executives to prison terms. In addition, a consortium of shippers filed a class action lawsuit against the three shipping companies and received $52 million.

Just a few years later, TOTE's competitor, Horizon Shipping (owned by the huge international private equity firm Carlyle Group), was socked with another whistleblower case: the shipping company had fired John Loftus, a captain with nearly four decades of experience, for expressing concern about the seaworthiness of Horizon's ships. Loftus won his case and was awarded $1 million. In early 2015, Horizon went belly-up.

TOTE's steamships were now carrying nearly 30 percent more cargo.

After losing its price-fixing lawsuit, TOTE restructured the company to squeeze more money out of its operation. Critical onshore staff were designated redundant and let go, along with some of the company's most experienced seamen. As office staff was shuffled around and new corporate entities sprouted from the morass, communication unraveled. It was often unclear who was in charge of what as managers and employees were moved, given new titles, or let go. Critical emails concerning safety or staffing aboard the ships went to a dizzying array of people, and sometimes went unanswered because it was unclear who was supposed to respond. What emerged was a pass-the-buck culture.

In 2012, TOTE scrapped *El Morro*, replaced it with *El Faro* (which had been laid up since 2006 in the Port of Baltimore), and put in an order for two new LNG-powered vessels to take over the Puerto Rican run. Ashore, staff attention shifted from the two steamships at sea to the new pair coming online and the daunting task of creating an LNG infrastructure to support them.

By the fall of 2015, TOTE Maritime and Crowley held a virtual monopoly on Puerto Rico's lifeline. How could TOTE continue to distinguish itself when vying for the lucrative Walmart and CVS Puerto Rican shipping contracts? Reliability. On-time delivery. Fast turnaround. *El Faro* and *El Yunque* were fast—sailing on average 20 knots. It was an easy sell to customers. Former Horizon captains tell me that TOTE ships always got to San Juan first. For many retailers, speed and on-time arrival were worth premium pricing.

Meanwhile, *El Faro* and *El Yunque* served their company, inspected but, nonetheless, in disrepair. Though the ships were well built and made of thick steel, decades of use had eaten away at their integrity. Some problems were apparent. Those things got fixed. Some weaknesses—rot and rust from life spent in the brine—lurked out of view in dark recesses. Those things didn't get fixed.

Keeping ancient machinery operational requires constant vigilance. The crew did its best to keep the old ships running. But it was hard to get in front of the issues. On her final voyage, *El Faro* was loaded down with cargo, more than she'd ever carried, her rusty decks riding dangerously close to the water.

CHAPTER 16

DAWN

The tumbling and heaving of a ship in a storm can throw you right out of bed. Back and forth, from one side to the other, your body's weight goes with the ship. You lose your sense of up; the world becomes a sickening swirl of constant, unpredictable motion. There is no center. The stomach revolts and you begin to retch. Your organs churn as violently as the sea.

As *El Faro* steamed deeper into Joaquin's grip, the off-duty seamen braced themselves in the corners of their cabins wearing their life vests while trying to sleep.

That night, as *El Faro* drew closer to Joaquin, Chief Mate Steve Shultz could not have slept well.

He unsteadily made his way up to the bridge at 3:44 a.m., dodging from grab bar to grab bar, timing his steps with the ship's roll, and when he finally got up there, he wanted to know what the hell was going on.

He saw the form of Second Mate Danielle, her face illuminated only by the dim lights on the console. He saw the shadowy form of helmsman Jackie Jones standing by the ship's wheel. And he felt a succession of riotous waves knocking them off course with every hit. The autopilot alarm sounded regularly—it couldn't maintain its programmed bearing. Shultz wanted to steer the ship by hand, but it was difficult to turn into the waves when you couldn't see them.

Shultz was the first deck officer to notice that the ship was listing. An inclinometer on the bridge —a curved clear tube filled with liquid, like a carpenter's level—told him that as she rolled, the ship was dipping farther over to starboard than port. She wasn't properly righting herself. He attributed the vessel's list to the strong port winds. But without the anemometer, he couldn't be sure.

It was certainly roaring out there, though. How much wind did it take to send them that far over to one side? He didn't know. He'd only been on *El Faro* a few weeks. Must be wind heel, he concluded, and made incremental heading changes to try to maintain their 116-degree course.

AB Frank Hamm lumbered up to the bridge for his watch. Leaning as the floor rose and dipped beneath him, gripping the grab bar on the console, he surveyed the scene. "Captain ain't been up yet?" The question hung in the air like a black cloud.

A few minutes later, Davidson came through the bridge door. He didn't like his crew's sullen mood. "There's nothing bad about this ride," he told them. "I've been sleeping like a baby."

Shultz was done with heroics. "Not me."

"What? Who's not sleepin' good? How come?" Davidson demanded.

"I could tell from 1:00 a.m. that the storm was comin' in," Frank told the master. "Those seas are for real."

"Well, this is every day in Alaska. This is what it's like," Davidson declared.

Shultz immediately stepped up to support his commanding

officer. "That's what I said when I walked up here. This is every day in Alaska."

"A typical winter day in Alaska. We're not pounding."

"Exactly my words," said Shultz. "This is every day." But that wasn't exactly true. While storms in the Pacific Northwest could be savage, they didn't compare to the highly organized cyclonic Atlantic hurricanes.

Davidson went on, "I mean, we're not even rolling. We're not even pitching. We're not pounding."

Shultz wasn't so sure. "We're heading up into the gusts and coming back down, riding through a lot of rain here."

"Yeah, I can see that," said Davidson. "Steer it up. Just steer in this direction right now."

"Yes, sir," Shultz said.

Like Shultz, Davidson considered the effects of wind heel—a list to starboard caused by strong winds blowing against the broad port side of *El Faro*. The crew could counter the list by shifting water from the starboard ballast tank to the portside ramp tank. It wouldn't make a huge difference, but it was something. He called down to the second assistant engineer to find out how they were doing down in the engine room. His engineer confirmed that all was well.

But lack of experience was beginning to show. Former officers of *El Faro* who rode the ship in heavy seas told me that when she rolled over for an extended period of time, the engineers would always call up to the bridge to complain, and with good reason. Solids stirred up by the ship's motion would enter the fuel lines and clog them up. Either someone in the engine room was furiously monitoring the fuel strainers, or no one was paying attention. If an engineer was struggling down there to keep the lines open, the bridge should have been alerted.

El Faro was equipped with bilge alarms in each hold that rang an alarm and lit up a light on the console in the engine room. It

would have taken about eighteen hundred gallons of water to ac-
tivate a high-level bilge alarm in a level condition. We know that
the alarms had been recently tested. Is it possible that they mal-
functioned? Or that they were on the port side—the high side—of
the ship? Or worse, that an inexperienced engineer had turned on
the bilge pumps (which may or may not have instantly clogged)
and never gave them another thought?

The third assistant engineer was twenty-six-year-old Mitch
Kuflik. Another third assistant engineer, Michael Holland, was
only twenty-five years old. He'd been shipping for three years and
spent only half of that time on a ship; he had a lot to learn about
the complex steam engine. The last third assistant, Dylan Meklin,
was a fresh graduate of Maine Maritime. He'd just stepped on
El Faro on September 29 at age twenty-three. This was his first
voyage, his first real job, and likely his first tour on a steam ship.

In the engine room, two huge boilers—fire chambers where
water is turned into steam—sat opposite a ten-foot-long console,
built in the 1970s, along with the rest of the ship. To young guys
like Holland, Kuflik, and Meklin, raised in a digital world, this
console must have seemed like a nightmarish vision of an ana-
logue past: buttons, lights, gauges, and levers. Alarms were always
going off. Things were always going wrong. Some were critical,
some could be ignored. Less experienced engineers depended on
the chief engineer's knowledge for guidance.

The second engineer, Keith Griffin, came on watch at four
o'clock in the morning. Following protocol, he diverted a portion
of the engine's steam from the propulsion system to clear accumulated soot off the boiler tubes. This is called "blowing tubes."

Engineers I spoke to after the sinking find this troubling. The
ship was very close to the eye of Joaquin. Containers were already
shaking loose from their lashings. Inside, it must have felt like a
carnival ride with all that rolling. This wasn't the time to perform

routine maintenance, especially when it compromised the ship's power and speed. GPS records show that blowing tubes that morning slowed the ship down three knots for about thirty minutes.

In an interview, Chief Engineer Rich Pusatere's wife said that the daily demands of *El Faro*'s ancient systems had left him exhausted. Someone needed to wake him up, if he wasn't in the engine room already.

On the bridge, Davidson continued to ignore or even deliberately avoid evidence that might contradict his stance. "It's probably better off that we can't see anything, Chief Mate," he told Shultz.

Davidson went downstairs to see how the galley was faring in the storm, maybe get some breakfast. As soon as he was gone, Shultz took charge. He used the ship's internal phone to check in with the second assistant engineer again. Something didn't feel right.

"It should go without saying the weather decks are secured," he told the young engineer, "but I want to just make sure you wrote it down because the third engineer came up from the second deck when he got off watch and we don't want anyone else doing that." There was an interior route from the engine room to the ship's house, but some people preferred to walk up the loading ramp, through the watertight door at the top, and onto second deck, which by this point was awash in waves.

For the third time, Shultz assumed that the scuttles had been properly closed without checking on them himself. The barometric pressure on the ship was down to 960 millibars; Joaquin's eye measured 950 millibars with maximum sustained winds of 121 miles per hour. *El Faro* was steaming straight in.

At 4:45 Davidson downloaded the BVS report that had been waiting for him since eleven o'clock the night before. At that point, the forecast was more than fifteen hours older than the most recent NHC report; it was completely obsolete. The BVS forecast continued positioning Joaquin's eye much farther north, giving Davidson false reassurance that he'd successfully steamed below it.

When Davidson connected to the internet to get his BVS report, he also uploaded any of the crew's emails that were waiting in a queue for the satellite connection. That's when Danielle's final three messages went to her mother and two friends. The one to Laurie read: "Not sure if you have been following the weather at all but there is a hurricane out here and we are heading straight into it. Category 3 last we checked. Winds are super bad and seas are not great. Love to everyone. -D."

Several hours later when her mother woke up and saw that message, she knew all was lost. In the decade that Danielle had been shipping as a deck officer, Danielle never told her mom about the dicey situations she'd occasionally find herself in until after she'd been through them. And she never wrote "love."

On the bridge, bad news began flooding in from around the ship. Containers and trailers were coming loose and falling off the deck, crew reported from below. One shipping box equipped with a GPS locator dropped over the side and drifted south between the islands, following the very course Danielle had charted for *El Faro*.

A few minutes later, the bridge phone rang again. It was the chief engineer, Rich Pusatere.

He wanted to know what they were doing about the list. The ship was heeling over so far, Pusatere said, that he was having trouble keeping his lube oil pressure up.

The lube system bathed the huge turbines and the propeller shaft in highly refined oil to prevent friction. All the oil ran through a closed system of pipes and gears and collected in a sump—a large shallow pan—below the huge bull gear. The mouth of the system's suction pipe, much like the plastic tube on a liquid soap container, sat twenty-three inches right of the centerline of the ship, ten inches above the base of the pan. If oil levels were high, the mouth would remain deep in oil. But with all the ship's pitching and rolling, the oil sloshed away from the mouth, causing the pump to suck air.

At home, when you're low on liquid soap, the tube attached to the pump hovers above the liquid. You can pump all you want, but it will never draw soap. Either refill the bottle, or throw it away.

That's exactly the situation Pusatere was facing. He had a main pump and a backup pump if the first one failed, but neither would work if the suction mouth wasn't making contact with the lube oil. What's worse, the backup pump hadn't been working well on prior voyages according to Jimmy Robinson's turnover notes. It had been scheduled to be rebuilt.

If both pumps failed, Pusatere had a gravity tank located high above the engine room loaded with reserve oil. By opening up a valve, he could get about nine minutes' worth of lubrication to flow down into his turbines before they seized completely.

His lube oil problem was compounded by the fact that *El Faro* left Jacksonville on September 29 with low oil levels—at 1,225 gallons, about 200 gallons below the recommended operating volume. One former *El Faro* chief engineer testified that he always ran the system at higher levels to prevent loss of lube oil suction, especially when he anticipated foul weather.

If Pusatere lost his lube oil system, he'd lose the turbines. They'd shut down almost immediately. If he lost the turbines, they'd have no way to steer *El Faro* through the hurricane. She'd drift, turn her broad side to the seas, take on water, and eventually succumb to the ocean. The solution was simple: they had to steer her into a smoother ride, get her vertical.

Davidson listened to Pusatere's concerns. Severe winds on his port side must be blowing them over more than usual, he told him, as he'd told his officers. He couldn't see the waves in the dark and with all the ship's movement, it was hard to physically judge exactly how much the ship was hanging as she rolled from side to side, but he could feel it.

That slow movement indicated flooding. But all the howling and rocking in the black night were fatal distractions.

CHAPTER 17

THE RAGING SEA

23.50°N -73.85°W

At 5:00 a.m. *El Faro* was drowning. An angry sea raged on, joining forces with the power of Joaquin's eye wall to form a thirty-five-mile-wide, aqueous tornado. Lightning shattered the darkness, turning torrents of rain whipping across the ship's windshield into bright white claws. Furious gusts made a deafening howl on the bridge. The ship jerked and plunged as though she had lost her mind with fear.

In the oppressive roar, Captain Davidson planted his feet at the chart table, gripped the grab bars, and once again began rationalizing conflicting weather reports. BVS and NHC put Joaquin's eye at different points on his map. Was he north of the storm? In it? South of it? Desperate for an explanation, he blamed the discrepancy on the failings of modern technology.

"Here's the thing," Davidson told his chief mate. "You've got five GPSs on this ship, so you're gonna get five different positions. I use one weather program—BVS."

Davidson refused to consider the idea that the BVS report was simply wrong. It told him where the ship was relative to the hurricane's path calculated more than nine hours before, but Davidson dismissed anything that countered his faith in it, including the NHC's reports, Weather Underground, and his own colleagues. He couldn't face the fact that every single decision he'd made since the beginning of the voyage had put his ship in dire straits. And that the course change he'd commanded Danielle to make at two o'clock in the morning—without consulting his weather reports—had sealed *El Faro*'s fate.

What kind of man is capable of facing such excessive errors of judgment?

Shultz knew where they were relative to the storm. He'd been keeping a close watch on the barometer. He saw that it was damn low, 960 millibars, indicating that they were frighteningly close to Joaquin's eye. In the final hours, he would monitor that barometer as if it were his secret Cassandra, cursed to tell him the truth even when his captain refused to believe.

Instead of retreating, Davidson dug in. He killed the autopilot and called out steering commands to his helmsman, Frank Hamm, peering into the vast darkness as if staring down a mortal enemy, trying to anticipate its next swing.

Midships. Rudder left ten. Right twenty.

Now this was taking command.

"Hard right," Davidson told Frank Hamm.

"Rudder is hard right," Frank confirmed.

"Our biggest enemy here right now is we can't see," Davidson called to Frank with growing frustration. Then he announced to all on the navigation bridge: "We're on the backside of the storm," even though there was no indication that that was true.

Once again, Davidson turned to Shultz for assurance that this weather wasn't all that bad. "You get seventy days of this shit up there in Alaska," Shultz dutifully parroted.

Everything loose on decks below was slapping wildly in

the gales; containers yawned over the decks as if weighing the odds, then plunged overboard and floated away. Jeff Mathias had watched all this from down below. He fought his way against the bucking ship up to the bridge to find out what the hell they were doing about it.

Up there in the tower, the winds were louder. The night was blacker. The raging sea, cloaked in darkness, was chilling in its ferocity. Flashes of lightning illuminated a brutal watery world of monstrous waves and seafoam. Jeff waited for his eyes to adjust to the dim light, then studied the captain with incredulity. Finally, he yelled out, "What's the wind speed?" Davidson didn't answer. Maybe he didn't hear. Mathias asked again.

"We don't know," the captain shouted back. "We don't have an anemometer."

Mathias was a mariner as well as a farmer. At forty-two years old, he'd spent his life learning about man's machines and man's crops, and the colossal damage Mother Nature could inflict on both. He was knowledgeable and intuitive and he knew how a ship should feel. This one wasn't right. When she rolled, she wasn't coming back up to center.

"Never seen it hang like that before," he yelled at Davidson.

"Never? We certainly have the sail area to cause a list like this," Davidson told him, once more blaming the ship's pronounced lean—up to 18 degrees—on the wind blowing square against the port side of the ship.

As *El Faro* rolled to starboard and lingered there, Mathias wondered aloud about the chief engineer, Rich Pusatere, no doubt toiling deep down in the ship's engine room, trying to keep the steam engine going. Davidson told Mathias that the engineer was having trouble with lube oil pressure. More alarm bells went off in his head. As a chief engineer, he knew that loss of lube pressure spelled doom.

Jeff's concern spurred Davidson to adjust their course once

again. He told Shultz to change *El Faro*'s heading from 100 degrees (zero is north, 90 is due east) to 60 degrees (northeast), closer to the wind. Davidson assumed things would shift around, but maybe getting her upright would help his chief engineer address his oil problem.

In fact, it was a huge change. Instead of running parallel to the waves, the ship would be nearly perpendicular to them. At that point, instead of rolling, *El Faro* would pitch. She would muscle her way up every swell, hang for a moment over the crest, then race down into the trough like a roller coaster, her bow crashing into the sea when she reached the bottom, causing a magnificent splash.

Pitching would cause new problems for the engineers: Every time the ship tipped into a dive, the propeller would come out of the water. Without resistance on the blades, the screw would spin wildly, radically changing the load on the engine. If this speed went unchecked, the force of the shaft spinning that fast would cause catastrophic engine failure. To prevent his overspeed relay from tripping (which would immediately kill the propulsion system), Rich had to gear down the shaft's RPMs by hand each time the stern came out of the drink.

The officers on the bridge could only guess how high the waves were. Winds at 100 knots form thirty-foot seas. But that's only an average. Every third or fourth wave could be fifty feet. They were steaming their way up a mountain of water and rolling into deep troughs. As dawn broke, they would be blinded by the waves created by the bow as it crashed into the brine.

Any minute now, Davidson thought, the wind will shift, confirming they're south of Joaquin. But he was wrong. *El Faro* was deep in the grip of an eye wall so powerful that no engine on Earth could pull the ship out. The barometer plummeted to 951 millibars; they were right next to the eye.

With the ship rising, falling, and slamming in the darkness, it felt like a free fall each time they surfed down a wave. At the wheel, Frank Hamm was pale and stricken. His fear was palpable. Davidson turned his attention to his panicky AB.

"Mr. Hamm," Davidson advised, "stand up, hold on to that handle. Just relax. Everything's gonna be fine. Good to go, buddy. You're good to go."

"You all right?" Shultz asked the AB with compassion.

"How you doing, Frank?" Davidson chimed in. "You want a cup of coffee?"

"Yes."

"How do you like it? Black?"

"Just black."

"You want the chair, Frank?" Davidson asked.

"Yeah, yeah. Yes, sir." The big man sat down at the wheel, no longer concealing his fear. This was the worst weather he'd ever been in.

Davidson bravely stared down Joaquin, trying to ease Frank's panic. "It sounds so much worse up here on the bridge," Davidson said. "Down below, it's just a lullaby. It's only gonna get better from here," he announced to the troops. "We're on the backside of it."

But they weren't.

At 5:43, Rich Pusatere called to the bridge with more bad news.

It wasn't just wind heel that had sent *El Faro* leaning far over to starboard, he told them.

They were flooding.

The scuttle on the weather deck that led to three-hold—the ship's largest hold—had been left open. It might have been left ajar during loading. It might have been closed but poorly secured. It might have been so banged up over time that it could never have been properly sealed in a storm like that. Regardless, every wave that washed over the deck sent an ocean through the deck opening into the bowels of the ship. The water was collecting on the starboard side, where she was already leaning due to wind heel, adding extra weight that pulled *El Faro* farther over, preventing her from properly righting herself. She wasn't coming back up. Jeff Mathias was right.

Water was also coming into the engine room through a door leading to three-hold that was supposed to be watertight.

"We've got a problem," Davidson said. He sent Shultz down to the second deck with a flashlight to find out what was going on. "On your way down there, knock on Mike Holland's door." Rich needed all hands in the engine room to man the pumps, and he had requested his third assistant engineer.

Shultz headed for the bridge door when Rich called up again. "Where's the water coming from?" Shultz asked him over the phone. The captain took the receiver, listened, and repeated what the chief engineer told him: "Bilge pump running. Water rising."

Davidson instructed Rich to move ballast from starboard to port to try to right the ship. A minute later, he decided to turn the bow farther to the north. They were now at close haul to the wind. He hoped that this change would help center the vessel, giving his crew a chance to see what was going on down below.

Davidson commanded Frank to steer them into a heading of fifty degrees.

"Left twenty," he told Frank.

"Left twenty," Frank repeated.

Things weren't getting better, Rich told Davidson. If the starboard list was such a problem, the captain figured, then he should turn the ship even farther to get the wind on the other side. "I'll get it going in that direction and get everything on the starboard side to give us a port list and then we'll be able to get a better look," he told his chief engineer.

Davidson's solution was overly simplistic. True, if they turned just a few more clicks more to the west, Joaquin's powerful northeast winds would catch *El Faro* on her starboard bow and throw the whole ship over to port. But such a maneuver poorly timed could cause the ship to instantly capsize if the bow dives in the wind and a wave crashes onto the deck, sending her into a catastrophic roll. Anyone who has tried to handle a sailboat in huge gales knows that when you tack—point your bow to the other side of the wind—it's very violent. The power of the sail's boom when the wind catches it and it flies across the keel can knock a person out.

El Faro didn't have a boom, but she did have cars, trucks, and containers straining against their chains and lashings.

"Left twenty," Davidson told Frank.

"Left twenty," Frank repeated. They were turning the mammoth ship into a heading of 350 degrees—north-northwest— exposing their starboard side to the storm.

Remembering that he'd tried to ballast his way out, Davidson grabbed the ship's phone to reverse his commands. "Stop transferring the ballast from starboard to port," Davidson ordered his third engineer.

Danielle came up to the bridge at six o'clock. It wasn't time for her watch, but she couldn't sleep through the storm. When she saw their new heading, she was astonished. "Three-five-zero!" she cried out. Why had they changed course? Were they heading back to Jacksonville?

"Hi," Davidson said to her casually.

"Hi," she said, echoing his apparent calm. "How are you, Captain?"

"How are you? A scuttle popped open and there's a little bit of water in three-hold. They're pumping it out right now. Chief's down there with Jeff Mathias. They're closing the scuttle." No big deal, he implied.

Rich called up to the bridge with more alarming news. Water wasn't just coming through the scuttle. It was coming through the ventilation trunks. The ocean was pouring through the louvered vents above his engine room and flooding everything. It could short out his console.

The lube pressure wasn't coming back, either, in spite of their course change.

Davidson confirmed that they were bilging as much as possible, then turned to steering the ship, reversing his commands.

"Right twenty," Davidson commanded, sending them back through the wind.

"Rudder right twenty," Frank repeated.

Shultz called up to the bridge on the handheld radio. He was on the second deck with his flashlight, knee-deep in water, trying to steady himself while looking through the scuttle above the three-hold. With every slam of *El Faro*'s bow, however, water splashed out from the hold over the edge of the open hatch. The massive cargo space must have been heavily flooded. Thousands of gallons of water had somehow gotten into the belly of the ship. It wasn't wind heel that had pinned her down to starboard. It was the weight of the water.

In the whipping winds and waves, Shultz secured the scuttle. But it was too late.

Danielle stood on the bridge next to Davidson, waiting for orders. When the captain talked about getting the latest BVS report, she quipped, "It's stormy." What else did he need to know? She was exhausted by nerves and fright.

"If you don't need me—" She was about to leave. But there was something about Davidson in that moment that appealed to her compassion.

"You want me to stay with you?" she asked him.

"Please."

"Down to 950 millibars," Davidson noted. That's astonishingly low. "Feel the pressure dropping in your ears just then? Feel that?"

Frank Hamm had his hands on the wheel, but he was visibly shaken.

"Take your time and relax. Don't worry about it," Davidson told the panicking man standing before him, gripping the small ship's wheel. They couldn't just pull over and wait it out.

"Stand up straight and relax," Davidson told him.

"I am relaxed, Captain," Frank replied.

"Relax. Steer the direction we're going."

At that moment, the wind finally caught the starboard side and violently threw the ship to port, along with everything she carried. Below, a tidal wave of water, cars, and trucks crashed against the inside of her port hull, pulling the ship into a deep roll. If *El Faro* capitulated far enough, it would exceed its stability and capsize.

The lube oil in the sump sloshed over to the opposite side of the pan, too far away from the suction mouth. Pusatere lost oil pressure completely. Without oil, the emergency trip shut off the turbines. Although the boilers continued to crank out steam, the giant shaft in the engine room came to a halt. Three minutes later, *El Faro* lost propulsion.

Davidson, Danielle, and Frank waited in shocked silence on the bridge. Time slowed to a crawl. The sky began to take shape. Dawn.

CHAPTER 18

WE'RE GONNA MAKE IT

Shultz, Davidson, Danielle, and Frank Hamm braced themselves on the bridge holding their breath, waiting in the howling dawn for word from the engine room. Without propulsion, they couldn't steer. Instead, the old ship lolled helplessly in the churn of the sea until her bow pointed longingly back to Jacksonville. For several agonizing minutes, Davidson stared at the ship's phone, hoping that Rich Pusatere would call with good news. In the engine room, the chief engineer was trying to get the turbines back online but water was sloshing around everywhere, pouring in from above, seeping through the watertight doors. He didn't have enough oil to get his lube pressure up.

Third Assistant Engineer Mike Holland was opening valves in the engine room manifold to bilge three-hold. It would take hours to pump out all that water. "Is there any way to tell if you actually have suction and it's pumping?" Davidson asked him over the ship's phone.

Not really. Unless someone could look over the side of the

ship and actually see the water flowing out of the hull, it was almost impossible to tell. There was even a chance that, as Holland opened valves, water got inadvertently pumped into other holds. Mariners who knew those ships told me that the combination bilge/ballast system aboard never worked well; the valves between the two intake and outtake systems were always getting pinned open by chips of rust or paint causing ballast water to flood the fourth decks.

High-water alarms were sounding in the engine room, but no one on *El Faro* knew where all the water was coming from or where it was going. The old watertight doors between holds began weeping; some of the man-doors inside those doors may have been left open during loading. Water rained down onto the engineers from the ventilation ducts in the ceiling and seeped through the doors protecting them from the holds. Soon every chamber in the ship would be filled with seawater.

Davidson ordered Holland to open all the valves on the bilge system to pump all the holds at once. But if some spaces they were pumping were dry, the entire system would gurgle and sputter to a halt.

In the end, it didn't matter. The ocean had found another way in.

Twenty feet below the waterline on the starboard side was the ship's main fire pump. It was connected by a short length of pipe to a small hole in *El Faro*'s hull. The pump was designed to draw seawater through piping to the engine room manifold. From there, engineers could open and close valves to direct water to extinguish fires anywhere on the ship. The crew regularly used this system to wash down the ship's decks, so the main valve between the fire pump and sea was nearly always left open.

As the cars in three-hold slipped their moorings and floated free, they crashed back and forth against the hull with every roll of the ship. Some cars probably slammed into the fire pump, busting the pipe, causing an eight-inch breach in the hull. At that depth, the pressure of the sea was tremendous. Using Bernoulli's equation, one small hole twenty feet below the waterline would look like a fire hydrant, blasting water at an astounding 161 gallons per second. Water exploded into *El Faro*, causing immediate and catastrophic flooding.

Rich Pusatere suspected this was happening but couldn't stop it. At their best, his bilge pumps moved sixteen gallons of water per second. No way they could keep up.

Davidson felt that *El Faro* was drowning and yet he waited to sound the general alarm. In that ferocious hurricane, they were far safer on a sinking ship than battling 100-mile-per-hour winds in *El Faro*'s open lifeboats.

He told Danielle to go downstairs and wake up Jeremie. She'd been groggy when she first arrived on the bridge, but now she was wide awake. On her way to his cabin, she changed into her work clothes. When she got back to the bridge, Davidson asked her to prepare an emergency message to shore. He assured her that it was just a precaution.

Danielle sat down at the communication desk and scrolled through the preprogrammed messages in the ship's system. Which one should she use? Flooding? Disabled and adrift? "I would do a bunch of them," Davidson told her. Who should she send it to? Davidson advised her to pick addresses off the provided list. He didn't mention anyone specific. She found an address for the coast guard and an email for TOTE.

At 6:39 a.m., the message was ready. She waited for Davidson to give her the word. "Don't send anything yet," he told her.

Frank Hamm noticed water dripping from the bridge ceiling. "Nah, that's okay," Davidson said. "Don't worry about it."

Without RPMs the rudder was useless. Frank had nothing to

do. He stood up and took a cup of coffee with Splenda. Now that they rolled with the waves instead of fighting them, it was a little easier to move around the ship.

Jeremie came up to the bridge as requested. "Am I relieving watch?" he anxiously asked. "Tell me what to do and I'll do it." The captain sent him downstairs to bang on doors.

"I want everybody up," Davidson told him.

Wind ripped a piece of handrail off the deck and sent it screaming by.

"On the *El Morro*," Danielle said, "we were on the same run and one of the hurricanes developed right over San Juan. We took a 38-degree roll. Oh my lord, that was annoying." She laughed brightly. "Imagine that." She was comforted by the fact that she'd lived to tell the tale. They'd be okay. Another great sea story.

Davidson picked up the satellite phone and called someone on shore. He left a message: "It's miserable right now. We got all the wind on the starboard side here. Now a scuttle was left open or popped open or whatever so we got some flooding down in three-hold—a significant amount. Everybody's safe right now. We're not gonna abandon ship. We're gonna stay with the ship. We are in dire straits right now. Okay, I'm gonna call the office and tell them. There's no need to ring the general alarm yet. We're not abandoning the ship. The engineers are trying to get the plant back, so we're working on it, okay?"

He hung up and once again declared that they were on the backside of the storm. Then he checked in with Rich; the engineers were still having trouble getting the engine back online, he told the captain, because of the list.

Davidson picked up his satellite phone again and called Captain John Lawrence, TOTE's designated person ashore. Again, he got a voice mail and left a message. At 7:00 a.m., Davidson called TOTE's emergency call center. He had a marine emergency. Could he please speak to a qualified individual?

Seven minutes later, Lawrence was on the phone. We have one side of that conversation:

Hello Captain Lawrence. This is Captain Davidson.

Sir.

Yeah, I'm real good. We have secured the source of the water coming into the vessel. A scuttle was blown open by the force of the water, perhaps. No one knows. Can't tell. It's since been closed. However, three-hold's got a considerable amount of water in it.

We have a very, very healthy port list.

The engineers cannot get lube oil pressure on the plant, therefore we've got no main engine. And let me give you a latitude and longitude. I just wanted to give you a heads up before I push that button.

Our position is latitude 23-degrees, 26.3 minutes north. Longitude 73-degrees, 51.6 west.

Yep, the crew is safe. Right now we're trying to save the ship. But all available hands.

We are 48 miles east of San Salvador.

We are taking every measure to take the list off. By that, I mean pump out that hold the best we can, but we are not gaining ground at this time.

Uh, right now it's a little hard to tell because all the wind is on that side too so we got a good wind heel going. But it's not getting any better. And I'm gonna guess, yeah, I'm, I'm—go ahead, sir. Go ahead.

All right, priorities: We're gonna stay with the ship. No one's panicking. Everybody's been made aware. Our safest bet is to stay with the ship during this particular time. The weather is ferocious out there. And we're gonna stay with the ship. Now as—go ahead, sir.

Right. The state of the weather: Swell is out of the northeast. A solid ten to twelve feet over spray. High winds. Very poor visibility.

Fifteen degrees, but a lot of that's with the wind heel.

I can't determine that at this time.

That is correct. The engine room has informed me that they are pumping that hold. There's a significant amount of water in there.

That's correct.

Yep, what I wanted to do. I wanna push that button, that SSAS button. I wanna send some alarms on our GMDSS console. I wanna wake everybody up.

Okay. I just wanted to give you that courtesy so you wouldn't be blindsided by it and have the opportunity. Everybody's safe right now. We're in survival mode right now.

Yep. Thank you, sir.

Davidson hung up the phone and told Danielle to sound the abandon ship alarm.

"Well, all hell's gonna break loose with the messaging and stuff like that," he told her.

Then suddenly, "Wake everybody up!" he commanded urgently and angrily. "Wake 'em up!"

"We're gonna be good. We're gonna make it," he announced to the void.

CHAPTER 19

WE'VE LOST COMMUNICATION

At 6:30 in the morning on October 1, Petty Officer Matthew Chancery was finishing his morning coffee. He was fresh off a three-day break, ready to start a twelve-hour shift monitoring distress calls from vessels in the 1.8 million square nautical miles of ocean known as District 7. It's the US Coast Guard's busiest area, encompassing coastal waters from South Carolina, Georgia, Florida, and the Bahamas, all the way down to Puerto Rico. About two-thirds of all coast guard search and rescues happen here.

Chancery had started his career in the Marine Corps; when his father fell ill, he began looking for other ways to serve his country. After talking to recruiters in various military divisions, he found himself drawn to the coast guard. He liked their mission of helping, not hurting. It's something you hear a lot when talking to coasties: Many served in the air force or the navy before switching over to the coast guard mid-career. Not that the other services were solely focused on killing, but they spent a lot of training time preparing for warlike scenarios that may never come, which can get tiresome. Most coasties get to use the skills they've acquired, if not daily, at least frequently. There's always a drug interdiction, or a missing vessel, or someone abandoning ship. And in late summer, there were always hurricanes.

Nearly everyone in the coast guard has at least one great rescue story. You could casually talk to someone like Commander Charlotte Pittman, a former national champion rower, now running media relations out of the US Coast Guard headquarters in DC. Give her an hour or two and she'll reveal that she'd been a coast guard helicopter pilot in Alaska where, on moonless or stormy nights, it was so dark that it took every bit of her concentration to avoid hitting the six-thousand-foot-high mountains all around.

One night on a rescue mission to help a disabled boy in a remote village, the wind was blowing so hard that rain washed across Charlotte's windscreen sideways. We're used to watching rain falling down, so either her instruments were right or her eyes were right. Charlotte had to battle her instincts not to pitch the bird to get the rain going the way her brain told her it should. Two emergency techs in the back of the helicopter had no idea what mental gymnastics their pilot was going through, and their confidence in her kept Charlotte focused. After a while, though, she couldn't fight off her impulses anymore.

She turned to her copilot, who was older and more experienced, and said, "I'm fucked up. Can you take over?" He did. After a few minutes, Charlotte noticed their airspeed dropping. "In training, they tell us that when you start losing your grip on reality, you physically pull into yourself," she says, holding an imaginary control stick above her lap and pulling it closer into her body. The copilot was unconsciously slowing the helicopter into a hover. She turned to him and said, "Hey, are you okay?" He answered, "I'm fucked up too." She says the only way they got home was by working together, talking each other through it, and remembering the sick boy in the back of the helicopter.

At the US Coast Guard's D-7 Search and Rescue Command Center in downtown Miami, Chancery's days could be just as psychologically harrowing. He regularly got calls from cruise ships requesting medevac for passengers with medical emergencies—often heart attacks. Frequently, a boat's EPIRB (emergency position-indicating radio beacon) went off, requiring the coast guard to determine whether or not there was a true emergency. Occasionally, Chancery's team got a call from a boater who spotted an emergency flare. In that case, he would send out a helicopter or two to find out what was going on. There were vessel collisions and sinkings. Everything—every single call that came into the command center—was investigated using the coast guard's regional assets: a handful of helicopters, boats, and planes stationed around the area.

Coordinating a search and rescue from an over-air-conditioned, windowless room can be emotionally draining. Just two months before, on July 24, a couple of fourteen-year-old boys disappeared off the coast of Florida in a nineteen-foot, single-engine boat, triggering the largest search-and-rescue effort the coast guard had executed in recent memory. Chancery's team had nothing to go on except reports that the boys left from Jupiter, Florida, and may have been heading to the Bahamas.

Over the course of a week, dozens of vessels and air support scanned more than sixty-six thousand square miles hunting for the missing teens. Finding a boat without an initial position takes patience, especially so close to the powerful Gulf Stream, which has a tendency to pull things north around and down again, like a giant horseshoe. Chancery tried to calculate the possible drift of a vessel about the size of the one lost. Eventually, the boys' capsized boat and a life jacket were found dozens of miles off the Florida coast far north of where they'd departed, but Perry Cohen and Austin Stephanos were never found.

Chancery arrived at the command center at 5:30 a.m. on October 1. He lived about twenty miles away from the downtown headquarters, but at that time of day, traffic was light. Getting to work on his Ninja wasn't a problem.

Nestled among Miami's brand-new skyscrapers, hotels, and condos, the coast guard's concrete bunker of a building sits like a can of Bud among fine wines. It's easy to miss the entrance, protected by an armed guard and a metal detector. Deep within, in a single room built for machines, not humans, the command center's white walls and white floors are bathed in fluorescent light and the hum of air-conditioning. It's a fortress, accessible only to those with the highest security clearance. Drug interdiction is a big problem in D-7, and one person always mans the law enforcement station, watching the computers and phones. On one wall, a giant grid of LCD screens show maps of the area, constantly updating with weather and vessel information.

That morning, Chancery and three other operations specialists on the day shift met with the guys coming off the night shift to get a sense of what was happening out there. They didn't have much to talk about, since it'd been a quiet night, but they did discuss the weather—there was a hurricane idling a few hundred miles east of the Bahamas, which meant that the coast guard's helicopters on Great Inagua Island (between Turks and Caicos and Cuba) and Andros Island (in the Bahamas) would be out of play if something came up that day. From the most recent forecasts, Chancery and his team knew that the storm was predicted to cut north. Maybe it would eventually hit land.

As soon as Chancery logged into the system and settled into his chair, he got an urgent email from the coast guard's Atlantic Area Command Center, the central emergency station on the East Coast, located in Portsmouth, Virginia. The lieutenant commander who'd sent it followed up with a phone call to make sure someone had seen it.

Davidson had called Lawrence at seven o'clock that morning, warning TOTE's man ashore that he was going to trigger an Inmarsat C alert message—a manually controlled distress call that sends out a ship's location via satellite. Sure enough, that alert came into the Virginia command center about ten minutes later, at 7:13 a.m. from latitude 23°28'N, 73°48'W. The ship was traveling 235 degrees at 8 knots, approximately a third her normal speed.

The lieutenant commander told Chancery that he'd received two distress alerts—an Inmarsat C alert and an SSAS alert from a 737-foot ship in the Bahamas. The lieutenant commander added that he'd just gotten off the phone with a Captain John Lawrence—the designated person ashore for the ship's company in Jacksonville, a company named TOTE. Lawrence had called ahead of the emergency alerts to let the coast guard know that he'd been in communication with the ship's captain, and it wasn't an emergency.

Lawrence said that over the crackly satellite connection, the shipmaster's tone was calm. He'd assured Lawrence that the crew had everything under control. He said that they'd been taking on water through an open scuttle, had a fifteen-degree list, and had lost propulsion. But they were dewatering. Everything would be okay—they were simply setting off their alarms in case they needed help later. They wanted to make sure folks knew where they were.

In his retelling, Chancery makes sure I understand that in spite of this information, Lawrence told him that *El Faro* was in no danger of sinking. They discussed arranging for a tow, and Lawrence was calling salvage companies in the area to see if anyone was available.

The follow-up email from the Virginia-based lieutenant came shortly thereafter. It read:

> *"El Faro, San Salvador. We received the following Inmarsat C distress alert from motor vessel El Faro in position. MISLE reports it is 737-foot vessel. The vessel also activated its SSAS*

alarm. The command security officer for the vessel contacted the
vessel and relayed their condition to us. POC John Lawrence.
A scuttle opened in rough weather and the vessel took on water
creating a 15 degree list. They stopped the ingress of water
they are not at risk of sinking, but they are without power and
engines. They are dewatering the vessel. Please assume SMC
and work with RCC Bahamas to respond and report back with
your efforts."

(The US Coast Guard never misses an opportunity to use an acronym. MISLE is short for Marine Information for Safety and Law Enforcement, the coast guard's vessel and marine incident database; SSAS is the Ship Security Alert System, the emergency signaling system; POC is Point of Contact; SMC means Search and Rescue Mission Coordinator; and RCC is the Rescue Coordination Center.)

Chancery didn't know that the ship was American, but he used a database to look up its name: *El Faro*.

Chancery's first move was to plug the separate distress signals—the Ship Security Alert System (SSAS) and the Inmarsat alert—into his Search and Rescue Operations software on his desktop computer. It took him a few minutes. The alerts were about five nautical miles apart near a small, uninhabited island in the Bahamas called Samana Cay.

Emergency alerts go from ships to satellites and back to Earth. Where they ultimately get picked up depends on which satellite gets them first. In *El Faro*'s case, the distress alert went to a Norwegian satellite, picked up at a station in Norway. From Norway, the message was sent to Virginia, but the message was repackaged before it was sent. Between Scandinavia and America, some critical information was lost, including the ship's course, speed, and time of reported position, which, by the time Chancery got it, was outdated.

Chancery's training had taught him that secondhand information isn't as good as first. He wanted to talk directly to the ship, so he needed the vessel's satellite phone number. He called the person listed in his database under *El Faro*, a Captain John Lawrence, the same man who'd talked to his colleague in Virginia a few minutes earlier. "Hey, John," Chancery said. "I'm calling you back. You were listed as the point of contact for *El Faro*."

"That's correct," Lawrence answered.

"Do you have contact or direct communications with the vessel?"

"I did. They called me. I was just trying to call them back. And the satellite dropped the call. I can give you the phone number."

"Yes, give me the phone number to the vessel. That's fine."

"Okay. Satellite number—you have to dial 8-011 first to get the satellite," said Lawrence, who then passed along the information.

"All right," answered Chancery.

"That's what he called me on," said Lawrence. "And I tried calling him back a few minutes ago to see if they had had any contact with you guys yet."

"They haven't."

"Yeah, I talked to the other coast guard guy. Can you tell me what the plans are now? He said you were going to contact the Bahamas, I guess?"

"So here's the deal," Chancery said. "That depends. Right now, based off all the information you've provided, I'm not in a distress phase currently because they said they're not at risk of sinking and they have dewatered. And I see they are without power and engines."

"Correct."

"Are they able to anchor that boat right there?"

"They're forty-eight miles east of San Salvador, so I don't think so," Lawrence said.

"But the position that I'm looking at says they should be able to anchor," Chancery said.

"Oh, really?"

"It's not that deep. And they're near some small islands."

"You have a better map than me," said Lawrence with a laugh. "I'm sorry, your last name was?"

"Chancery. And right now, yes, I am going to pass this information on to the Bahamas and you know, how we handle this depends on the situation because this is a large motor vessel."

After he hung up with Lawrence, Chancery tried to call the ship more than a dozen times. "I was just dialing, dialing, dialing, dialing," he says. "*All right*, I thought. *I got to get a hold of this ship.*"

Everything changed when Chancery got an EPIRB alert from *El Faro*. In his mind, this signaled that she was in a grave situation. "I was already moving to, *There's something seriously wrong here*. If this captain had this happen, and now he's not answering the phone, now I've lost communications with a mariner that's in distress. We needed to get something moving here."

Usually when a vessel sends out a distress alert, its location goes with it. "An EPIRB alert marks the boat right on the map," Chancery says. "Boom! Here's where the EPIRB is." But *El Faro*'s emergency alert beacon wasn't GPS enabled. It was an older model that hadn't been updated since it was installed nearly a decade earlier. Instead, it eked out a few distress pings between 7:36 a.m. and 7:59 a.m.—read as a "406 unlocated alert"—which the coast guard picked up, but couldn't effectively use.

To accurately pinpoint *El Faro*'s position, low earth-orbiting satellites would have had to pass above the ship while its EPIRB was pinging so that they could use the Doppler frequency shift to estimate its location. In the twenty-four critical minutes that *El Faro*'s EPIRB pinged and then vanished forever, no satellites were in range. This would prove one of the most frustrating hurdles in the massive search-and-rescue mission to come. No one knew exactly where the ship was.

With growing urgency, Chancery reached out to his operations team in the Bahamas and Turks and Caicos, where a total of three MH-60 Jayhawk midrange helicopters plus an MH-65 Dolphin short-range helicopter were stationed on Andros and Great Inagua Islands. He already knew their answer, but he had to try. He said to the aircraft commander, "Man, I got something going on. If I need you, can I get you guys to go in?" The pilot answered, "No, we can't even pull the bird out of the hangar." At this point, the storm was getting close. There was no way they were sending their helicopters out in that.

Chancery turned back to his computer to try to find vessels close to *El Faro*. Maybe someone was out there who could help. Nearly every ship has an automatic identification system (AIS) on board that regularly reports its position, via satellites, to parties that monitor marine traffic. Websites like vesselfinder.com make this information available to anyone with a computer. The coast guard can access AIS data to get both the ships' positions and contact information. In this case, Chancery turned on everything to search hundreds of miles around *El Faro*'s last known position. He identified a few vessels in the area and tried to contact them one by one. He finally got hold of the *Emerald Express*, the only boat even close, about sixty miles from where he thought *El Faro* might be.

He said, "Hey, what's your weather like out there?"

They were sheltering in the lee of the southernmost island in the Bahamas and were seeing 80-mile-per-hour winds, thirty- to forty-foot seas. "I want you to do callouts for the SS *El Faro*. I'm looking for them," Chancery told them. "Here's my phone number. Call me back if you establish communications with them." They never did.

The *Emerald Express* was having problems of its own. Joaquin was pounding the small Panamanian-flagged barge with twin engines too weak to overcome the force of the storm. A crew member of the *Emerald Express* later told a reporter, "The winds just threw us where it wanted to. We almost rolled two or three times and we lost one empty container. That's when we started to prepare life jackets to abandon ship." But they hung on, tossed around like a cork, and finally made it into a shallow inlet so flooded at the time that the crew thought they were still in deep water. When the flooding abated, they were shocked to discover that they'd been grounded twenty-one miles inland on tiny Crooked Island.

Chancery had run out of options. Inside the command center, it looked like just another day—the other officers were doing their thing, hunched over their computers. But Chancery began to feel panicky. *Oh my God, something's, something's really wrong here*, he said to himself. Big American container ships don't just vanish. He went to his commanding officer, Captain Coggeshall, for guidance.

Coggeshall had been a coast guard pilot for twenty-some-odd years. He'd seen a lot and knew people. He replied, "Hey, NOAA has these Hurricane Hunter C-130s"—large, fixed-wing aircraft sturdy enough to fly into the eye of a storm. See if you can get them on the phone. Call whoever you need to."

In his three years at D-7, Chancery had never reached out to NOAA before to request the use of one of their planes. But he was willing to try. Because when it came to search and rescue, no one had ever said no to him before. *And, obviously, since it's a C-130 aircraft*, he thought, *they can't land, they can't pick anybody up out of the water, and they can't hoist. But they're there. They're over it*. He asked NOAA, "Can you make callouts? Can you get a hold of the ship? Can we find out what's going on?"

But you can't just fly into a Category 4 hurricane and look for a ship. Visibility is next to zero. You need coordinates. Otherwise, where were they supposed to look? Calculating *El Faro*'s position

proved near impossible. Chancery had been given three conflicting data points—the ship's AIS, the location from the SASS, the
Inmarsat coordinates—all collected at different times. Which one
should he work off of?

And if the ship *had* lost propulsion as the captain said it did,
then the vessel was drifting, but in which direction and at what
rate? Chancery tried building a drift model like the one the coast
guard had used to try to find the two boys off the coast of Florida.
But to get quality output, he'd have to overcome several hurdles.
For one thing, the modeling software he was using maxed out at
50-mile-per-hour winds, and he had reports that the winds out
there were at least 80, maybe 100.

Worse, the software was designed to model vessels a third
of *El Faro's* size. Then things went from bad to awful. The coast
guard had just upgraded its entire modeling software system a few
weeks before. They were still working out the bugs on October 1,
but hadn't provided their operators the option to switch over to the
old system in the event that the new one failed.

The coast guard's servers couldn't keep up with the calculations
Chancery was inputting, and they started crashing. He had to reboot his computer again and again, losing data every time. Either
the case got corrupted or was unrecoverable. Later on, the entire
program and sometimes even the entire workstation would reboot
or close out mid-scenario. Chancery would spend twenty minutes
calculating a drift scenario, and then "it would just be gone."

The tools he'd been given to do his job were useless. But Chancery is the kind of guy who had to do something. He is built big
and looks like a classic marine, with a buzz cut and a cleft chin.
If he wasn't chained to that desk, you could easily imagine him
jumping into a boat himself and heading into the storm.

Captain Coggeshall, Chancery's commanding officer, was
confused by Lawrence's assertion that Davidson was calm and
had things under control. Maybe the vessel was disabled, maybe

its antennae had been knocked off, which is why they couldn't communicate with the ship. But Coggeshall may have also known about the hydrophones that had picked up a giant thud on the ocean floor near *El Faro*'s last known position. Then again, though the hurricane was intense, Davidson had said that all was well. It didn't add up.

From his coast guard training, Coggeshall knew the basics of search planning the old-fashioned way. In desperation, he and Chancery turned to pen and paper, drawing lines on a map and calculating vectors. Objects drift within a 45-degree vector downwind. "So I used the last known position of *El Faro* as my minimum," Coggeshall explained. "I figured that the hurricane was going to be the biggest factor—that ocean currents were not going to be the driving force for the drift of the vessel or the life rafts or lifeboat. So I basically ran the drift at 40 knots, multiplied it by three, and applied it in a northeast direction, which was basically the direction the storm came from and was heading out again. I figured wherever the storm was going it'd pull anything on the surface with it."

It was the only thing they could do to keep moving forward, because they'd eventually have to tell their planes, helicopters, and cutters something.

CHAPTER 20

SEARCH AND RESCUE

Great Inagua, the second largest island in the Bahamas, is home to mosquitoes, donkeys, and a gleaming white $20 million US Coast Guard hangar. The previous hangar was flattened in 2008 by Hurricane Ike, whose 145-mile-per-hour winds ripped off its roof and bent its massive steel I-beams like paperclips. The new hangar is designed to withstand a Category 3 hurricane. Above its high-strength precast concrete panels, a thick web of steel trusses keeps the roof on during extreme updrafts. On one side, four sixty-thousand-pound motorized doors slide open so the coast guard's Jayhawks can be towed in and out.

All of District 7's aircrew serve a two-week stint on the south-ernmost Bahamian island each quarter. Most of them don't mind being out there—staying up late, cooking steaks on the open grill, playing games on Xbox. For a handful of men and women in their thirties, it's about as close as you can get to sleepaway camp. Except every twenty-four hours or so, you get to fly the coast guard's bright orange-and-white $30 million Sikorsky birds.

On September 15, 2015, MH-60T rescue pilot Lieutenant Dave McCarthy deployed to Great Inagua. (Coasties call it GI.) Dave had never been there before, but he'd heard about it from his colleagues. With his Roman profile and small blue eyes, Dave looks ice cold until he breaks into a warm, generous grin, which, fortunately, he wears most of the time. Exactly one decade ago, he arrived at the doorstep of the US Coast Guard's Tampa Bay recruiting station after a rocky career in technology. He didn't know what he was looking for. He'd just come back from a trip to DC, and after visiting the Pentagon, the memorials, and Arlington Cemetery, Dave found himself "truly moved" to help his country. He told the recruiter, "I just want to serve." After signing up for the reserves, he started to think more seriously about a military career. Helicopter pilot came up as an option. His grandfather had been a pilot in World War II, almost an ace. When Dave told his dad, who'd served in the air force during the Vietnam War, that he was thinking about becoming an aviator, his dad said, *Make sure you like flying first.* So he paid for a short helicopter ride around Disney "just to see if I enjoyed it," Dave says. "And I was like, yeah, this is pretty cool."

Dave was deploying to GI with fellow pilot Lieutenant Rick Post, a God-loving Christian from Nebraska with a boyish face, big brown eyes, and hard-to-miss jug ears. Rick grew up hunting and fishing, driving tractors and chasing cows, "doing all the Nebraska country boy stuff," he says. He'd always wanted to fly jets. He applied to the Air Force Academy but, he says, "I guess the good Lord had other plans." He didn't get in, but he did get accepted to the Coast Guard Academy. "So I was, like, do they have an aviation program? Yes? Okay, I'll do it."

Rick and Dave geared up for two weeks on GI like modern-day Robinson Crusoes—night-vision goggles, flight suits, and forty pounds of plastic-wrapped raw steak and chicken packed in coolers. They boarded the C-130 in Clearwater and flew across

Florida, past Miami, and then over open water. On their right, they could glimpse the coast of Cuba. They flew over the hundreds of islands that make up the Bahamas, and when they reached GI, the plane did an "ass pass" to clear the donkeys off the runway before touching down.

After two uneventful weeks out there, they were preparing to head back with their crew to Florida when on Tuesday, September 29, Clearwater called to tell them that they were staying put. A hurricane was coming; they might need extra hands out there. The C-130 from the Air Station landed in GI a day later, but not to take them home. Instead, it carried an additional aircrew. District 7 was staffing up for something big.

Back at Air Station Clearwater, Florida, Captain Rich Lorenzen was hedging his bets. As one of the commanding officers responsible for the coast guard's air assets in D-7, he worked with his fellow officers to shuffle helicopters and flight crews among the air bases like chess pieces. He'd served eight years in the army and twenty-five years as a coast guard pilot and operations officer in District 7's hurricane-prone territory and, in that time, learned that being prepared always paid off. He'd overseen the evacuation of GI ahead of Hurricane Ike, a huge effort that's still a point of pride for the career military man: "We got eight C-130 sorties to empty that place out before it hit. I call it the greatest coast guard airlift story that was never told."

Now Lorenzen was watching Tropical Depression 11 and didn't like what it was doing. It defied expectations— a very bad sign. "Usually depressions that start up in the higher latitudes zigzag up and then go off Bermuda and you never see them again," he says. "This was up there meandering around. And all of a sudden, it started growing. And I'll never forget—it was on Monday morning. We came in and we're having our morning briefing. And I was thinking, *We gotta keep an eye on that one.*

"So I got together with a small group of folks after the meeting and said, 'Hey, we've got the swap-out at GI happening tomorrow. Instead of pulling those folks out, I have a feeling this storm could get bigger, based on the potential track. How about we leave those folks where they are and bring in additional crews? So if this turns into a big storm we're not caught saying, Man, I wish I had crews there.'

"And sure as heck, it started to blossom into a hurricane pretty quick and then took the southerly jog straight toward our bases."

Stranded on GI, Rick and Dave, along with the rest of the crew, spent Wednesday afternoon prepping the base for hurricane-force winds in case Joaquin went right over them, like Ike did seven years before. They tied down anything that a storm could send rolling or flying. They filled up buckets with water to flush the toilets, and others with drinking water. Once everything was secured, Rick and a few other guys drove around the island in the truck the coast guard keeps down there. It was a gray stormy day. To clock wind speed, they drove with the wind at their backs until it stopped pushing against their outstretched hands—40 miles per hour. Where they could really tell that there was a hurricane off-shore was by watching the ocean. "It was just churning, heaving, roiling," Rick says. "It was way higher than normal, tearing up the boat ramp, putting rocks up on the road."

At the island's lighthouse, where the Bahamian waters are usually a calm cerulean, waves crashed halfway up the tower. "Poor little crabs were clinging onto the rocks for dear life, trying not to get swept away," Rick remembers. "It was definitely the sea state that tells you there's a hurricane going on."

Thursday morning, they got the call directly from District 7 command center in Miami: an American ship called *El Faro* was heading through the storm. It was having communications issues. Could they fly over and check it out?

Mission requests came from District 7, but all missions had to be cleared by Clearwater's expert operations officers who work with the on-site pilots to assess risk and either accept or decline. This mission presented insurmountable challenges. In Clearwater, Commander Scott Phy examined the radar with the pilots on GI—*El Faro* looked like it was dead center in the eye of a Category 4 hurricane with sustained winds up to 125 miles per hour. The Jayhawks could get close—maybe within fifty miles or so, but that wouldn't help anyone. They'd have to wait until the storm edged off.

Part of that risk assessment with Clearwater included discussion about the ambiguity of *El Faro*'s situation. "We didn't have good reports," says Commander Phy, a decorated pilot who was working with Lorenzen as the operations officer that day. "D-7 was talking to *El Faro* through the shipping agent who said that they had a couple of mechanical casualties but they were okay. So at first, it just seemed like they were—from the reports we were getting—having some mechanical failures, but they were okay."

Lorenzen, the commanding officer, remembers it the same way: "The first real *El Faro* report I got was during one of those hurricane phone conferences with the command center, maybe in the afternoon," he says. "And the message was, *Hey, they seem to be doing okay. And I remember personally thinking, Whoa. Why is this thing heading through that storm? That's not a good idea. And I'm starting to wonder if they had communication problems. Do they not have the ability to look at weather graphics or talk back to their home base or to their ownership and see that this storm is brewing and they're heading toward it?"*

He and Phy waited for more information from the command center, and slowly, the full extent of *El Faro*'s situation emerged. "There were a lot of hours of uncertainty when a lot of us just assumed, *Okay, they're probably going to get to ride this out. Hopefully they turned away, but maybe they've lost communications because they're in the middle of a hurricane. So nobody can communicate. Got it. That doesn't mean they've sunk. It just means nobody can talk to them.* But when I heard that the ship's captain had said they'd lost propulsion and had flooding, I knew things were going bad. Once those containers start shifting over the side of the thing, you get too much weight on one side, and with twenty-, thirty-, forty-foot seas hitting that thing, anything can happen. I knew it probably got ugly real quick."

The pilots on GI told Chancery that there was no way they could reach *El Faro* in their Jayhawks that day and declined the mission. So he spent Thursday working with the sketchy information he'd been given by TOTE's designated person ashore, John Lawrence, and tried in vain to establish communications with *El Faro.* He also struggled to create a drift model for a search-and-rescue plan working with the conflicting position data he'd received. Was the ship actually in trouble? Though his instincts told him yes, there was nothing else he could do. Joaquin had whipped itself into a frenzy and wasn't budging from where *El Faro* theoretically was.

As evening approached, the officers in Clearwater agreed to send a C-130 to *El Faro's* last known position by first light Friday morning.

Rick and Dave were on standby on GI. They headed to bed early, assuming they'd fly out at dawn the following day to work with the C-130 to find *El Faro.*

Around 9:00 at night, the phone rang. It was D-7. But the call wasn't about *El Faro.* A 212-foot Bolivian-flagged container ship named the *Minouche* had been brutally battered by Joaquin and lost her engines. The *Minouche* was listing thirty degrees, and the twelve people aboard were preparing to abandon ship.

For the second time that day, the pilots on GI discussed flying into Joaquin with their commanding officers in Clearwater. The *Minouche* was farther south than *El Faro,* so conditions would be slightly better, but flying at night is always challenging. The visibility ceiling out there was low, about three hundred to four hundred feet high, and the winds were in excess of 40 knots. That meant they'd have to fly close to the water if they wanted to see anything. If they hit a downdraft, they wouldn't have far to fall before they got tangled up in the ocean's waves.

Dave, Rick, rescue swimmer Aviation Survival Technician Ben Cournia, and flight mechanic Aviation Maintenance Technician Josh Andrews would be flying under extreme conditions at night, hoisting survivors one by one into the Jayhawk. Josh would man the hoist and call out positioning commands to his pilots up front. It was a dangerous, delicate operation, requiring a symphony of perfectly timed maneuvers.

Dave and Rick had never flown in a hurricane before. But there were twelve people in the water. That's twelve lives they could save. "That's pretty significant gain, right?" Dave said to me later. After a brief risk assessment over the phone with Commander Phy, the crew agreed to take the mission.

They had thirty minutes to get airborne. The crew zipped into their drab olive flight suits, grabbed their helmets and night-vision goggles, and headed to the hangar as the mechanics rolled back its massive doors and towed the bird onto the tarmac into the howling storm. The flight crew climbed in and flicked on the lights. In spite of the winds, mosquitoes were swarming—thousands of black specks appeared across the windshield and lurked inside. Once strapped in, the four men could barely move; they were being eaten alive right through their suits. They quickly started up the engine, scaring off the mosquitoes with the machine's vibrations, got the rotors going, and lifted up into the black of night.

Rain washed sideways across the windshield as 80-mile-per-hour gusts knocked them around. Visibility was zero. Jayhawks always shook, but in those conditions, it was bumpier, like driving down a rocky road at high speed, Dave says. Even with their night-vision goggles, Dave and Rick couldn't see anything in the darkness, so they flew by instruments. Before they took off, Dave was concerned that his radar would pick up too much noise to reliably warn him of hazards on land, so he used his computer mouse to input the outline of Inagua as flight points on his screen; that helped him follow the island's coastline to ensure they didn't hit anything like buildings or antennas. It would also help them find the unlit runway on their return.

They went slowly to get the hang of flying in those conditions, taking almost half an hour to reach the *Minouche*—just forty miles south of their base.

When they arrived at the distressed vessel, they saw a ship glowing brightly in a black sea, listing wildly to port. They flew a wide circle around the scene to get a sense of what was going on. The *Minouche* was a cheaply built foreign-flagged ship not known for its seaworthiness; under the pounding waves, its hull had cracked in half. Luckily, its crew had good lifesaving equipment. Their life jackets, equipped with personal locator beacons, madly pinged the Jayhawk's sensors, making it easy for Rick and Dave to get properly positioned and strategize the rescue.

A big cargo ship and a Coast Guard cutter lingered a couple of miles away, but seas were too heavy for them to make any attempt to save the crew. If the rescue vessels got too close, one big wave could launch them right on top of the raft, crushing the people huddled inside.

On their first pass around the *Minouche*, Dave and Rick saw water washing up on her as waves battered her broadside. The small rubber life raft with twelve men bobbed nearby. During their second lap around, the aircrew pulled together a game plan—they'd go into a hover, lower Ben in, and have him pick the guys up out of the raft one by one. They had continuously trained for this scenario, but never under such crazy conditions. It was baptism by fire.

Rick threw a few flares into the water to help set up a visual horizon for himself, but the ocean immediately swallowed them.

At that point, water was halfway up the *Minouche*'s deck as the sea pulled her deeper in.

On the third lap around, they prepared to lower their rescue swimmer when Rick looked down. He couldn't believe his eyes. "Do you guys see what I'm seeing? Am I seeing this right?" he said in horror. The *Minouche*'s lights were still on, but she was completely under water. Getting smaller and smaller, vanishing into the deep. It was one of the eeriest things Rick had ever seen: watching a big ship sink before his eyes. Dave lifted his goggles to get a better look. But she was gone.

Wearing his neoprene wetsuit, mask, flippers, and snorkel, Ben buckled on his life vest, attached himself to the hoist, and crouched at the open door, watching the roiling seas below twist and heave in the bright white of the helicopter's spotlight. He took a minute to time the waves. His plan was to drop on the backside of a large wave, which would give him a few moments to orient himself before the next wave hit.

Josh braced himself against the chopper's doorway, blocking Ben from jumping, while calling out steering commands to Rick over the radio in his helmet—*forward right 15, forward right 10, forward right 5, easy, hold*—easing them into position above the raft.

Rick and Dave sat too far forward in the Jayhawk to see the hoist and raft directly below them, so they relied on Josh to be their eyes and ears. When they were in a good place, Josh gave Ben the signal—go time. The rescue swimmer gauged the rhythm of the waves, aimed for a trough, took a deep breath, and plunged flippers first, arms tight to his body, into the blackness. Josh quickly fed out the hoist cable, taking care not to jerk his swimmer, which could seriously injure him. Once during training, a rescue swimmer had two teeth knocked out that way.

Ben was one of the few people in his class to get certified as an elite coast guard rescue swimmer. The punishing Rescue Swimmer Training Program is notorious for its higher than 50 percent attrition rate. Along with intense physical training, the program demands superhuman ability to overcome mental and physical limitations. At one point, swimmers are pushed to extreme physical limits and nearly drown.

But all that training couldn't prepare Ben for the conditions in which he found himself that night, in the darkness with hurricane-force winds and huge seas. He plunged down into the warm, frothy waters, which felt womblike in the darkness and silence of the deep. As soon as he resurfaced, though, chaos reigned supreme. The Jayhawk's spinning blades above caused a powerful downdraft that competed with crosswinds and torrential rains to push him back under. The sound of the waves was deafening. He repressurized his face mask, and then braced himself for the next wave, which he knew was there but couldn't see in spite of the spotlights. The air was half sea spray, making it nearly impossible for him to clear his snorkel, and the ocean's churn knocked him around like a gang of thugs, trying to confuse and drown him. It was enough to make him wonder why he'd signed up for this. He used all his strength to keep his head above water, but the intense rising and falling of the swells made his stomach churn.

"I was telling myself this was for real this time—twelve lives were on the line," he said. "All our training came down to that moment."

Ben rode up the mountainous crest of a wave, set his sights on the raft, and began a strong, rapid crawl down into a valley, up again, occasionally glimpsing the weak lights coming from the life raft, gasping for breath in the soul-sucking wind. His body was pounded by ferocious waves, but he refused to give in.

A few minutes later, he drew close to the raft and grabbed a tight hold to a line. He shoved his head inside and saw twelve petrified men looking back at him. The crew was mostly Haitian and didn't speak English—eyes wide, dark black faces. But they understood the universal language of a rescue. The *Minouche* went down so fast that her crew didn't even know she'd sunk.

In the roar of the sea, Ben could yell but he wouldn't be heard. Instead, he pointed to one panicked seaman in the raft and signaled for him to climb into the rescue cage dangling above them. The mariner didn't want to leave the safety of the raft. Ben had to physically draw out the frightened crewman, wrestle him into the basket, then use hand signals to Josh to start hoisting. When the basket got to the helicopter's open door, the seaman inside clung to it like a crab in a trap, forcing Josh to shake him loose. The man hit the helicopter's deck facedown, and lay there, spread-eagle, frozen in shock.

Ben watched as the helicopter swallowed the basket, then he turned to hoist the next guy. But the raft was gone. In the short time it took him to load the first man into the basket, the raft had drifted a few hundred yards away. He fought his way through the waves to the raft. If he had to swim that far every time, he'd exhaust himself after a few hoists. Instead, he helped seven more guys up to the hovering Jayhawk, riding the helicopter's cable back to the raft between each hoist.

Dave had been keeping an eye on fuel, and now it was getting low. "We can't keep going," he told his crew. "We've got to go back to base to refuel." Ben stayed with the raft and watched the chopper disappear into the night. Though the water was warm, the wave action was sickening. He and the four remaining men, eyes wide as saucers, waited to the sound of the deafening storm, trying not to puke.

As they were pulling into Inagua with their rescues, Dave and Rick saw a lone bird standing on the tarmac, disoriented by the storm. "Aw, no, no, no!" Rick said, anticipating what would happen next. Like a rocket, the bird took flight and was instantly consumed by the helicopter's fifty-foot blades. Bird strike.

While the *Minouche*'s crew recovered in the hangar, the coast guard mechanics had to inspect the chopper to make sure the bird hadn't been ingested into the engine or otherwise damaged the aircraft.

With their rescue swimmer and four guys in a life raft battling high seas in the darkness, Rick and Dave didn't have the luxury of time. Josh and the line crew used a flashlight in the driving rain to pick bird parts out of the rotor and try to assess whether the Jayhawk was safe to fly. He did his best to clear it and ran his assessment by the captain in Clearwater. He got the go-ahead, and the aircrew took off, back into the night.

Ben was able to get the ninth guy into the basket when the flight mechanic noticed that the hoisting cable had frayed. They couldn't risk breaking the line, so once again, the Jayhawk was forced to head back to base. When they got there, due to a power failure, the hangar's sixty-thousand-pound doors were stuck closed. The mechanics, in desperation, pried open the hangar's doors with two small tractors called "mules," and towed out the second chopper.

But who would crew it? Piloting and hoisting require split-second decision-making based on a string of mental checks. Sleep deprivation degrades motor and processing skills. If they're really tired, pilots can forget things on their checklist and skip a critical step. Flying four or five hours on night-vision goggles only adds to their fatigue.

Josh, Rick, and Dave huddled around the phone and talked it out with their commanding officer in Clearwater. They knew they were tired—they'd backed one another up, tried to catch each other's mistakes as a team, tried to avoid micro-sleeps. Their adrenaline had spiked several times—when they first went out, when they were hoisting, when they hit the bird. But exhaustion could easily overcome any one of them, putting the mission at risk.

All that made it tougher to judge whether Phy should let the other crew take over.

Still, there were a few good reasons not to switch teams: Dave, Rick, Josh, and Ben had perfected their hoisting and communications system and were working well together. The guys in the raft knew the drill. And if the new crew went out, the twenty-four-hour-on-duty clock would start ticking for that team, limiting Inagua's ability to deal with anything else that came up during Joaquin, like a possible search and rescue for *El Faro*. Dave and his team wanted to finish the mission. They could do it.

They swapped helicopters, went back for the remaining three *Minouche* seamen, and returned to base at dawn on Friday morning. They'd been up for twenty-four hours.

And in that entire time, there had been only silence from the American container ship caught out there, somewhere in the storm.

CHAPTER 21

FLIGHT TO JACKSONVILLE

At 3:20 Thursday afternoon, October 1, Deb Roberts's cell phone rang. She almost didn't answer it. Deb was a school administrator in the tiny rural town of Jay, Maine, halfway between here and nowhere, hunting country supported by a big paper mill. She'd lived there all her life, raised her kids there. And usually when someone called from a number she didn't know, it was a sales call. Which is why she surprised herself when she answered it.

"Are you Mike Holland's mom?" the woman on the line asked. Oh yeah, it was a sales call. Deb's son Michael was a young mariner, just three years out of Maine Maritime Academy, and because he was away at sea so much, she took care of his finances. Her number was listed on his credit cards. "Yes," she answered warily.

"I'm calling from TOTE Services," the woman said. Deb's son was kind of a wild one. She wondered what kind of trouble he had gotten himself into now.

The woman identified herself as the crewing manager for TOTE. And then she said the one thing every mariner's mother never wants to hear: they'd lost communication with Mike's ship. It was in a hurricane.

Deb didn't know anyone at TOTE or on *El Faro*. She hadn't even known the name of the ship Mike was on until she got the phone call from the crewing manager.

Mike rarely talked about his life at sea, not because he didn't enjoy it. Just because when he was home, he was home. The sweltering engine rooms were a thousand miles away. Mike hunted for deer and rabbit, and he fished. He went to sea to make money that he could spend on his life back home.

When he was in high school, Mike wanted to join the marines. That scared Deb—the United States was in the middle of a couple of wars, and she didn't want her oldest son walking into that if she could help it. So she convinced him to consider Maine Maritime. One of seven maritime academies in the United States, Maine Maritime offered an engineering curriculum that you could use at sea or landside, running power plants. Mike's football coach at Jay High School had gone there; now he worked at the local paper mill. Deb thought it was a good compromise—Mike would have the structure he craved and a lucrative, vocational career—but ultimately, he had to decide for himself.

During the fall of Mike's senior year, Deb drove him 111 miles west to the school's campus on Maine's rocky coast. She hoped it would sell itself and convince Mike to give up his military dreams. It was the best she could do to keep her son safe in a dangerous world.

In many ways, rural Maine and coastal Maine are worlds apart. In and around Jay, things feel like they're coming to an end. Owners of the paper mill have been threatening to shut it down for years as demand for their product shrinks. In town and along the roads, there's not much to buy—used cars, guns and ammo, fishing gear.

When Mike found out that he could keep his shotgun at the

school and check it out when he had time off, or drop a line in the water and fish whenever he had a break, he knew he'd found his place. It helped that he loved engineering. Plus, Maine Maritime's football coach was showing interest in the five-foot, seven-inch linebacker who played like a three-hundred-pound bone cruncher.

That was seven years ago. Mike had been around the world on ships—in Thailand, through the Panama Canal, everywhere. He hated being away from his friends and family but loved traveling. That's why he was getting tired of his job at TOTE going back and forth between Jacksonville and Puerto Rico. He wanted some new scenery. The only reason he didn't look for another job was because he enjoyed his mates aboard the ship. They were like family to him.

Deb immediately thought the worst. She couldn't listen to the woman on the phone anymore. She called out for her coworker to bring a chair and braced herself as the woman continued.

There were lots of details that Deb didn't hear. But she did pick up something about a website that TOTE would be setting up to give hourly updates to families. "And then I hung up the phone and just lost it," she told me a few months later as we sat at her dining room table.

Deb's coworker offered to drive her home, but the only thing Deb wanted to do was talk to her husband, Robin. He was level-headed. He would chase away her panic. *These things happen, especially in storms,* he told her. *Everything will be okay. It's a big-ass ship.*

"And so I drove home and paced," Deb says. "And paced and paced. Went out and weeded the flower beds because it was early October. Raked leaves. I just had to keep busy because I just couldn't stand it. And I waited and waited. And it took a few hours for the website to come up, and then I checked it every hour. I checked it every hour all night long. I didn't sleep. Forever. And then it was Friday."

Deb didn't tell anyone about Mike, his ship, or the hurricane. She assumed the whole thing would blow over and she'd be able to go on with her life. No one questioned why she was out of the office that day. Weeks ago, she'd planned to celebrate the beginning of Maine's hunting season by trekking with her husband to their hunting camp for the weekend. Since Mike was away so much, she'd always borrow his shotgun and send him pictures of the birds she got with it. This time, she'd use her own, a Christmas gift from Mike. That's what she and Mike texted about on Monday night, the twenty-eighth, before he left Jacksonville. "You'll finally get to use that new gun I got you," Mike had written. "Yes," Deb texted back. "I'll try to get a bird for you. I hope your mom makes you proud."

She wasn't surprised that Mike didn't reply. Communication was never easy when he was shipping. He probably went to bed or got pulled into other things. She'd hear from him when he got to Puerto Rico on Friday.

But now Deb was sitting by the phone. Waiting. Checking TOTE's website. Refresh. Refresh. She found she could track ships over the web, too. She'd never tried before, but she located *El Faro*. "I could see the last time that there was a beacon, which made sense because they said they had lost communication with them. They couldn't track them anymore. So I was watching and I just kept checking and checking and checking and it was always the same last known location. Last known location."

In the morning, TOTE posted a new message on the website: the company was holding a phone conference for *El Faro* families in the afternoon.

Everyone desperately needed clarity. Some, like Jenn Mathias in her house on the southern coast of Massachusetts, tried not to think about the ship that carried her husband, Jeff, into the storm in the Bahamas. He'd been in bad weather before. There was one time when Jeff found himself in a North Atlantic storm near the place where the *Andrea Gail* of *The Perfect Storm* had gone down. The seas knocked around his vessel so hard that a forklift chained to the deck shook loose and crashed through the hull. Jenn showed me photos of the aftermath—it created a huge hole in the two-inch steel.

When Jeff first went to sea in 1998, he didn't have email, and cell service in those days was prohibitively expensive. He'd call to say literally *hello* and *good-bye*. Even so, his phone bill sometimes hit a thousand dollars a month. The couple could go that long without connecting. To be closer to her fiancé, Jenn occasionally traveled with him while he was working; she boarded the *Alaska* in San Francisco bound for Hawaii and spent the first few days on the ship sleeping, lulled by the gentle rocking.

When Jenn was pregnant with their first child, Hayden, Jeff knew he had to go for a chief engineer position to make enough money to support his family. He finally landed a job on SS *Great Land*, one of *El Faro*'s sister ships, built in 1975. He sailed as chief engineer on her until she was scrapped in 2011.

"He was so devastated because he loved the *Great Land*," Jenn told me as we sat in her kitchen, her two younger children, ages five and seven, occasionally joining us to look at photos of their dad.

Jeff's passion was steam. When he was at home, he'd spend hours on eBay hunting for vintage steam manuals, tools, and photos. He had amassed a huge collection of engineering manuals and textbooks from the late nineteenth and early twentieth centuries.

After losing the *Great Land* job, Jeff filled in as a chief engineer aboard TOTE steamships whenever someone needed to take a leave of absence. He was waiting for a permanent position to open up. When it finally did, Jeff was torn. Timing was bad—at that point, the kids were ages five, three, and eighteen months, and he didn't want to leave his wife alone with them for ten weeks at a time. So he passed on the opportunity and worked the cranberry bogs and pumpkin patches while inspecting ships locally and occasionally going to sea for shorter tours.

Jeff began overseeing the conversion of *El Faro* for the Alaska trade in February 2015. He continued on and off throughout the spring and summer. The pay was good, and the stints were short. He made sure he was home for his kids' first day of school, then went back to Jacksonville in late September.

Jenn remembers their final conversation. Jeff called on Tuesday night, September 28, before *El Faro* left Jacksonville. He wanted to Skype with the kids, but Jenn said no. "I told him, 'I've got them all bathed, all ready for bed. They're wound down. I don't want to get them all hyped up.'"

She sent Jeff an email on Thursday morning, October 1, to let him know that she and his mother had started hosting schoolchildren at the cranberry bog field trips, as they did every fall.

Jenn thought it was strange that she didn't hear back from her husband. Then the email bounced back—*undeliverable*—and kept bouncing back.

That afternoon, she got the call from the crewing manager at TOTE.

"The first thing that went through my mind was that Jeff had done something silly that got him in trouble," Jenn says.

The woman told Jenn that TOTE had lost communication with the ship. She assured Jenn that everybody was fine. Everything was fine. Jenn immediately called Jeff's best friend, Sean, who was also a chief engineer. Sean didn't answer. Then she went over to Jeff's parents' house, just a few hundred yards away from her home. The kids were watching TV. Barry, Jeff's father, didn't know what "lost communication" meant. But Jeff's brother did. "When I told him, his eyes bulged out of his head. Meanwhile, I'm thinking, everything is fine."

Sean called back later and assured Jenn that ships lose communication all the time. Maybe the storm had knocked out the satellite antennas.

In Jacksonville, thirty-year-old Marlena Porter had been tracking Joaquin for days. She couldn't help herself. She'd never felt completely comfortable when her husband, James, went to sea. He was in the family business—he'd been in the merchant marine for two decades, along with his father, mother, aunt, sister, and nephew. Regardless, the whole thing made Marlena nervous.

One time when James shipped to Africa, Marlena had a vision. She was with her mother in Miami overlooking Biscayne Bay and hadn't heard from her husband in a month. It hit her all at once. She looked at the water and had a vision of a ship sinking. The vision was so real that tears went running down her face.

What's wrong with you? her mother had asked.

She had said, *Nothing, nothing. It's just my allergies.* Marlena tried to pull herself together. But she looked again and saw the prow of a ship in the water. That's when her phone rang, a strange number. It was James. A month without hearing from him, a vision, and suddenly he calls. Was it a sign? *I am so scared for you to be out there*, Marlena told him.

I know, but I'm good, he had assured her. Every seasoned mariner could relate to her fear. But he was fine. *I don't want you to be worried*, he'd said. *Just make sure the boys are taken care of and stop worrying about me out here on this ship.* And then he reminded her, *I'm doing this to make sure we're good. Eventually I'll give it up. Eventually, because I do miss being home with y'all.*

Marlena had seen an opening. *I'll support you and stand by you the remainder of the way. But eventually you're going to have to come off that ship because I keep getting so many signs and it's scaring me.*

James said, *Don't do that. Don't do that.*

Recently, James had been doing a lot of planning, "getting things in order," Marlena says. He was making sure the kids were situated, making sure she had her degree. She found his behavior strange.

"So I took that as confirmation that God was telling me that it's time for me to prepare myself," Marlena says. "Build myself up and make sure I do everything I need to do to raise my kids, and do whatever it is to get everything squared away and situated. Everything that me and him discussed, everything that I can think about, it runs through my head every day. Because I can hear his voice, our conversation. It's just like he knew he was leaving."

One day in July, Marlena was in class when she missed a call from James. *Why didn't you answer the phone? I was trying to call you,* he'd said.

Well, I was taking a test, she told him.

You have to learn to answer the phone sometimes, baby, James had said, *because one day you may not be able to hear from me.*

James boarded *El Faro* in September and told his wife that he didn't know when he was coming home. The ship was going to be retired. James wasn't clear about what was next for him. *Are you going to be home for little James's birthday?* Marlena asked, because it was on Halloween, October 31. Her husband had answered, *I don't know. I just don't know. But I'll tell you this: If I don't make it there, to the birthday party, I need you to meet me in the Bahamas.*

Meet you in the Bahamas?

Yeah, that's where we gonna be at. We're gonna be in the Bahamas. She didn't understand what he meant by that.

Marlena's phone rang on the morning of October 1. She was working at Baptist Hospital, and the woman on the other end asked for her. *I'm calling from TOTE,* she said. *I just wanted to let you know that we were tracking* El Faro, *which is what your husband, James Porter, is on.*

Correct, Marlena answered.

We were tracking them and somehow they went into the eye of the hurricane.

"My heart literally dropped," Marlena recalls. "I went into a panic attack but I'm staring at this wall. And as I'm looking at this wall, it's like the vision that came to me before. I could see a ship going down. The first question that I asked her, I said, *'Did the ship sink?'* She said, *'No, no, no, no, no. No, Mrs. Porter, we're not gonna say that. We're not gonna speculate that the ship sunk.'*"

The woman had continued: *We're doing the best we can. We got everyone out there. They're going to search for them. We're going to make sure we bring them home safe and sound.*

"But I already had that feeling like he's not coming back," Marlena says. "He's not coming back."

When Laurie Bobillot got the call from TOTE, she assumed the worst, having already received that ominous email from her daughter, Danielle.

TOTE wasn't prepared to deal with the dozens of alarmed family members now demanding information about their husbands, wives, daughters, sons, and parents who were on *El Faro*. The first family conference call was "complete chaos," Deb Roberts remembers. "Everybody was just so upset and wanted answers. People were butting in and beeping in and out and they weren't muting their phones. It was just so frustrating."

But in that conversation, Deb learned that *El Faro* had lost propulsion. That the ship was listing at 15 degrees. And that the ship had been beset by Hurricane Joaquin. "I didn't know what listing meant. I didn't know what 'beset' meant."

During the phone call, TOTE assured family members that the company would pay for their travel and accommodations in Jacksonville while the coast guard searched for the ship. Many of the crew's families were based in Florida, but not Deb. "I'd already had it in my mind that my husband Robin and I were going because I just couldn't stand it. I couldn't stand being so far away, just so helpless."

She immediately bought plane tickets for herself and her husband. She got tickets for Mike's girlfriend, Kelsea, too. And she got a ticket for Kelsea's mother because she knew that she didn't have the emotional capacity to take care of the young woman who'd recently entered her life.

They booked tickets from Portland, Maine, to Jacksonville through JFK, but Deb still hoped for the best. "Honestly in my head I thought, *Okay, so this storm is clearing now. They'll be able to find the ship. By the time we get to Jacksonville, they'll know where they are. And Mike will come off that ship. And I will take him home.*"

When they got to Portland, they discovered that their plane was delayed due to communication issues.

Sitting in the airport, waiting for the plane, Deb saw a fiftysomething woman approach the counter. As she talked to the attendant, the woman broke down. Deb sat next to her husband, watching. The stranger's depth of frustration and desperation mirrored her own. This was an *El Faro* mother, just like her. She knew it. Deb followed the woman with her eyes as she walked slowly back over to the rocking chairs and stared out at the tarmac.

Deb walked over quietly and touched the woman's hand. "Did I overhear that you have someone on *El Faro*?" she asked.

"Yes," the woman answered. "My daughter."

"My son is on that ship."

Laurie Bobillot explained that she had moved to Wisconsin with her boyfriend a few years before and was trying to sell her Rockland house. Danielle, her daughter, a deck officer for TOTE, had been living there but shipped out half of the year and Laurie didn't think it made sense to keep the place and pay taxes on it. In September, Danielle had dutifully packed up her things. A few days ago, she'd flown back to Jacksonville to take over as second mate on *El Faro*.

Laurie stayed in Rockland after Danielle shipped to prep the house for real estate agents. Her daughter's two cats, Oprah and Spot, wandered around the remaining furniture and moving boxes, wondering what was going on.

When Laurie got the call from TOTE, everything that had happened in the past year came into sharp focus—Danielle's sense that her shipping career was coming to an end, her reluctance to get back on *El Faro*, her strangely emotional final email early that morning. There'd been so many signs. And now Laurie sat alone in the Portland airport, heading not to her boyfriend in Wisconsin, but to Jacksonville.

Watching the planes, waiting for some good news, Laurie looked up at the soft-spoken Maine woman standing next to her. The woman's face spoke of grief, fear, and hope—they were two mothers searching for their children. She instantly felt a deep sense of connection.

Deb and Laurie began working as a team. Worried that they'd miss their connecting flight in New York, they pleaded their case to the flight attendants who assured them that they'd call ahead and have the pilot hold the flight for them. The group sprinted like mad through JFK and arrived at the next gate just as the plane to Jacksonville was pulling out. That's when Laurie lost it, releasing the heartbreak raging inside her.

"You don't understand," she roared. "Our children are missing out in the middle of the ocean somewhere."

CHAPTER 22

SHIPS DON'T JUST DISAPPEAR

Clearwater Air Station commanding officer Rich Lorenzen had to make a career-defining decision. Should he send his aircrew into a Category 4 hurricane to search for *El Faro*?

Lorenzen had served in the military for three decades—including ten years with the US Coast Guard—and in 2015, was the top officer at Clearwater, one of the coast guard's biggest and busiest air stations. He was responsible for the welfare of more than five hundred pilots, rescue swimmers, and mechanics who provided air support to all of coastal Florida, the Gulf of Mexico, the Caribbean basin, and the Bahamas.

These were men and women programmed just like him—highly trained thrill seekers—whose jobs demanded that they balance their adrenaline addictions with unwavering discipline. Everything they do comes with risk, every decision has consequences. Every protocol comes with a checklist to assess danger each step of the way. The USCG motto is *Semper paratus*, always ready.

Most rescue pilots have the cool demeanor of race cars drivers—exquisitely sensitive to minute changes in their environment but highly directed in response. No action is wasted. Because they're in the business of saving lives, not taking their own.

By the time I met Lorenzen, he'd retired from the coast guard and was working as an emergency medevac pilot for Johns Hopkins All Children's Hospital in Clearwater, Florida. A fiftysomething guy looking every bit the part, he still sported his regulation haircut, now completely gray, and moved with the agility of an athlete half his age. When he met me, Lorenzen was wearing a flight suit on his taut frame. Flying never gets old for guys like him. Even with a storm brewing outside, he said he couldn't wait to strap himself into a helicopter, start her up, and take to the air.

Lorenzen led me to his small office on the top floor of the hospital where, as we talked, I watched the sky blacken through the window behind him. The gales came with a roar, pelting the rain obliquely as the rooftop antennas dodged and bowed. Turning around to admire the dramatic finish of a days-long storm system, Lorenzen's eyes positively sparkled.

Lorenzen was contacted by D-7 after *El Faro* had been out of communication for more than twelve hours. For all he knew, the ship could be adrift in that hellish weather, taking on water with every roll. If she was gone, her crew might be clinging to life rafts or bobbing in survival suits, battling every wave, fighting for air in the choking spray of the black, ferocious sea. Thirty-three people struggling in the darkness, praying for the bright white spotlight of a Jayhawk or US Coast Guard plane. Salvation.

As the commanding officer in Clearwater, Lorenzen's most important job was to weigh risk. To do this, he was part logistics expert, part cheerleader, and part psychologist. If even one crew member of *El Faro* was alive, he or she could still be saved. It had only been twelve hours and the seas were warm, reducing the risk of hypothermia—the way the majority of maritime victims die.

The will to survive is strong, Lorenzen tells me. He's seen it himself—people overcoming unimaginable conditions, tenaciously clinging to life for days, awaiting rescue. Every coastie has a litany of stories to back that up.

Commander Mike Odom, one the coast guard's first rescue swimmers, is a living embodiment of that.

On a stormy night in January 1995, Odom flew with his crew three hundred miles off the coast of Georgia to help three people trapped on a sailboat battling twenty-foot seas and gale-force winds. This was exactly what Odom had trained so many years for. He'd discovered his gift as a boy—swimming for hours every day as a mental escape from his tumultuous home life. When he was old enough, he left his rural Texas home and eventually joined the US Coast Guard.

By his twenties, Odom was a remarkable swimmer, strong enough to pass the coast guard's punishing eighteen-week rescue swimmer program.

That January night, Odom was lowered from the Jayhawk into the Atlantic. As soon as he unclipped himself from the cable, invisible two-story-high waves pushed him under the water. When he came up, choking on sea spray, Odom was completely disoriented. It took him a minute to get his bearings. Then came a flood of adrenaline. He cleared his face mask and snorkel and began his powerful crawl toward the vessel to hoist the three boaters, one at a time, into the rescue cage to safety.

Each hoist exhausted Odom. On the last hoist, the helicopter's pulley mechanism failed. Facing a three-hundred-mile trip back to land and quickly diminishing fuel, the flight team had to make an agonizing decision: leave now and abandon Odom, or try to get him back, risking everyone's life in the process. Those were the kinds of questions the crew had been trained to answer quickly. The answer was clear. Mario Vittone, Odom's flight team member and one of his closest friends, dropped a life raft down into the darkness, turned west, and headed to the mainland, eyes burning with hot tears.

Odom reached the raft and climbed in, but the churning sea kept tossing him out. Concerned that he would die of hypothermia and his body would be lost, he strapped himself to the float. Four hours later, he lost radio contact with the coast guard. His last message out: he was losing feeling in his legs; he was tired; he was cold.

A second helicopter crew arrived on scene about ninety minutes after that last message. When they focused their spotlight onto the bobbing life raft, Odom was motionless. They thought he was dead. A second rescue swimmer was dropped in the water and approached Odom; he shook and slapped him for about a minute, and Odom moved. He was severely hypothermic. But he was alive.

Now a youthful fifty, Odom travels the world inspecting ships for the coast guard. He keeps his dark silvery hair tightly cropped, but you can still make out a small patch of white in back—the mark of someone who has survived intense trauma.

The coast guard lore is rich with survival stories like these. Lorenzen told me about a FedEx driver who drifted in the ocean for more than seventy-two hours after his boat capsized off the coast of Florida. Thanks to his yellow shirt, he was eventually spotted by a sharp-eyed coast guard pilot and rescued. Steven Callahan, a sailor from Maine, overcame starvation, shark attacks, and dehydration to survive more than two months in an inflatable life raft, drifting eighteen hundred nautical miles from the Canary Islands to the Caribbean after his sailboat sunk. The coast guard had given up on him long before he washed up on the island of Marie Gallante.

These were the kinds of stories that sent otherwise normal, healthy people, with spouses and children at home, flying into hurricanes, diving into raging seas, searching for survivors.

The crew of *El Faro* might still be alive.

But this was no ordinary search mission. Hurricane Joaquin had continued along its excruciatingly slow and destructive path all the way to the Bahamas—flooding the coasts, tearing roofs off houses, and heaving the *Emerald Express* inland. Then Joaquin took a sharp hairpin turn out again. Seduced by the warm waters that birthed it, the storm went back east—almost directly over the last reported position of *El Faro*—to feed on the deep tropical seas, growing into a monstrous Category 4 hurricane, whipping the skies and seas of the Atlantic in its cyclonic frenzy.

Sending helicopters into that morass would be disastrous. Lorenzen's only choice was launching a C-130—a turboprop cargo and transport plane with a 132-foot wingspan—from Clearwater five hundred miles to *El Faro*'s last known position to see if they could spot the ship.

The coast guard doesn't normally launch its planes into hurricanes. The C-130 is the workhorse of the military, with a two-thousand-mile range, but its big airframe and giant wings make it unpredictable in extreme turbulence. C-130s cost up to $30 million new, so the coast guard gets the navy's hand-me-downs; they're coddled and cared for, Lorenzen says, but you just don't know how and when they're going to fail.

The aircraft definitely had limitations. Now Lorenzen needed to know his crew's. "So I sat down with [Lieutenant Commander] Jeff Hustace, one of the first pilots I was going to ask to go fly into this. I asked him, 'Hey, what are you comfortable with? What can the aircraft do? If you're comfortable launching, I'm going to ask you to try and get in.'"

Characteristically, Hustace said he'd try. People who join the aircrew of the US Coast Guard don't shy away from opportunities to test their skills.

Lorenzen's decision to green-light the mission was bolstered by his knowledge of storm systems. Flying into a hurricane isn't like in the movies, he says. You don't just hit a wall. As you move progressively through the outer bands of the hurricane, "the wind goes up 20 knots, then 30 knots. Ten minutes later it's up to 50 knots. Gradually that storm starts picking up as you get closer and closer." In other words, each band tests your skills as you work your way to the eye.

Lorenzen's most important function as a commander was to remind Hustace that he and his team had choices. If at some point Joaquin took charge, they could turn back: *If things start getting too hairy,* he told the flight team, *and you're getting tossed up and down, and there's lightning striking all around you, well, we don't want to lose a crew over this as well.*

That message helped Hustace and his crew make critical decisions for the next twelve hours. Their commanding officer trusted them to use their judgment first. His words would stay with them and eventually propel them deep into the storm.

Seven men—pilots, mechanics, and a dropmaster (trained to huddle inside the plane's open cargo door while airborne and drop flares, life rafts, and other critical supplies into the water below)—boarded the C-130 at Clearwater Air Station at 4 a.m. on October 2. It would take them several hours to reach the Bahamas at first light. Along the way, the crewmen psyched one another up for some hairy flying—the coordinates they'd been given put them right at the center of the storm.

As day broke, Hustace found himself in a blinding haze of thick clouds. At higher altitudes, Joaquin's rainbands—thunderstorms of increasing intensity that form concentric circles around the eye wall—forced him to drop lower so that his crew could get visual contact with the water's surface. He eased the C-130's yoke to settle the plane at twenty-five hundred feet above sea level as he nosed deeper toward Joaquin's core. Winds and rain rattled the aircraft; anything loose inside went flying.

And then they hit it: a huge, powerful downdraft caused by the low pressure between rainbands, which forced the C-130 into a near-instantaneous altitude loss of greater than nine hundred feet. If they'd been flying any lower, they'd have slammed into the water.

Hustace pulled them out of the dive and turned the plane away from the storm, then checked in with his team over the intercom. The crew was shaken up, but okay, and fully committed to the mission. They were looking for a huge American ship with thirty-three people aboard. That was a lot of potential survivors. To hell with a little turbulence. If there was anyone out there, if they could save even one person, they would. They'd come this far.

The C-130 made a high, wide arc out of the storm system and then took another pass straight into the cloud wall. Now Hustace had a better understanding of the storm's anatomy. Flying low, the crew searched below and made callouts into the void. The plane hit the downdraft again and plunged dangerously close to the roiling seas. But this time, Hustace was confident they'd make it through. As soon as he could, he pulled them out of the dive, climbed up to safety, and went in again. And again. And again.

With every pass, the crew searched the water to make out anything human from the milky churn. They were looking for a glimpse of red—the color of the thick neoprene survival suits issued to every mariner—orange life rafts, or the sharp edge of a lifeboat. (*El Faro*'s lifeboats were white, which hampered the chances that someone could spot them from the air.) The sea was a sickening riot of froth and foam billowing every which way, making it nearly impossible to make out something in the ever changing, tumbling landscape. Commander Scott Phy describes what it's like to try to search in those conditions, "You'll see a huge whitecap that looks like an overturned boat, all of your energy focuses in on that for a second, and then it vanishes." This happened to the air crew several times a minute, for hours and hours.

After a morning of brutal flying, Hustace's team refueled in Guantanamo Bay and headed back into Joaquin, but the C-130 was starting to show its age. Panels inside were shaking off, bolts were coming loose. The turbulence was tearing apart the old plane, and they were five hundred miles from home. Eventually, they were forced to head back to Florida.

Lorenzen says he'll never forget watching the C-130 land in Clearwater at 4:00 on Friday afternoon. As he walked onto the airstrip to greet his crew, he couldn't believe his eyes. Fuel was leaking out of the wing. "Literally that aircraft was getting jostled so hard that the turbulence caused the fuel line fittings to shake loose . . . I'm, like, *whoa, don't see that too often.*

"That's the extent of the bravery and the confidence that crew had, sticking their nose in to try and find those folks. Because for all we knew at that point, we were going to find thirty-three people. Either in onesies and twosies in survival suits. Or maybe, thank God, in lifeboats. And that's what we were waiting for, that's what we were looking for."

They got within sixty nautical miles of the storm's center—and found nothing of *El Faro.*

Meanwhile, the command center in Miami was charting out a search plan for the following day, hoping Joaquin would back off, affording them better visibility. They deployed three H-60 Jayhawks to Great Inagua from Clearwater. The coast guard also had two C-130s from Air Station Elizabeth City, North Carolina, standing by in Clearwater. Two cutters—the *Northland* and the *Resolute*—began to make their way toward *El Faro*'s last known position. They inched their way into the storm.

At 5:04 a.m. on October 3, the first US Coast Guard C-130 took off from Clearwater, arrived on-scene by 7:00, and tried to run the aerial search pattern Miami had laid out for them. But working with visibility as low as one nautical mile, they couldn't break through Joaquin's strong rainbands. The crew reported that the hurricane-force winds caused significant sea spray, whitecaps, and swells of twenty to fifty feet. They saw nothing as large as a container ship.

They were soon joined by an air force C-130 and a navy P8 jet.

As Joaquin inched farther east, away from *El Faro*'s last known position, a C-130 finally spotted a debris field in the water 120 nautical miles northeast of Crooked Island. The coast guard sent its Jayhawk in to take a closer look. The rescue swimmer dove into the frothy seas to retrieve three life rings, one stenciled *"El Faro."* It wasn't a good sign.

A second debris field ninety nautical miles northeast of Crooked Island revealed more detritus, presumably from the ship.

On day three, seven aircraft completed a total of twenty-eight hours of flight time searching an area of 30,581 square miles.

On day four, the weather cleared significantly with visibility at ten nautical miles, and winds down to 15 knots with two- to three-foot seas, meaning the cutters could finally get in there. The *Northland* arrived on-scene to take command of the search operation. Three C-130s, a navy P8 jet (a militarized Boeing 737), and a Jayhawk continued their hunt, along with three tugs chartered by TOTE, and the *Resolute*.

At 3:00 that afternoon, a rescue swimmer investigated a fiberglass lifeboat from *El Faro* first spotted by a Jayhawk. The boat had been crushed; its propeller blades bent and fouled. The crew of the navy P8 spotted two additional *El Faro* life rafts, one of which was picked up by the *Northland*. Both were empty.

Later that day, a twenty-seven-year-old rescue swimmer sat inside his Jayhawk watching a red immersion suit waving and bobbing in the water. Could there be a survivor in the suit? The swimmer had executed at least twenty-five rescues over his seven-year career and had flown a couple of search flights the day before. The weather was beautiful—warm waters, blue skies. They had been following their search pattern when his copilot had spotted the tiny red object. The pilot pulled the helicopter into a hover three hundred feet above the water as the swimmer clipped his harness to the helicopter's bear hook and had the flight mechanic lower him into the water. Once down, he swam over to the suit.

Within a few feet of it, the swimmer turned away in horror. Inside the red neoprene was a corpse. Floating for days in the warm salty waters, the head was triple normal size, "very bloated with a blueish skin tone," the swimmer told the NTSB. It was his first confrontation with death. He'd been trained to rescue the living. He took a moment to compose himself, then looked up toward the helicopter hovering above him and drew his hand across his throat to signal to his mechanic that the person inside the suit was dead.

They debated what to do. If they hoisted the body into the chopper, they'd have to return to base immediately and would be out of commission for a day. But what if there were other survivors nearby? Hovering above the body, discussing options, they got another call: a P8 had spotted a second survival suit waving in the water about forty nautical miles west-northwest of *El Faro*'s last known position. It could be a living person. They hoisted the rescue swimmer into the bird and dropped a buoy equipped with a GPS locator. The second survival suit was empty; wave action had lent it an eerie lifelike appearance. The beacon they'd dropped with the first suit failed to operate correctly and the body was lost to the sea.

Later, Captain Coggeshall of the coast guard considered what *El Faro*'s crew might have been facing. "Maybe they made it to the lifeboat or they had managed somehow to get a life raft in the water, but I couldn't figure out how in thirty- to forty-feet seas you safely get from the deck of the ship into a life raft that is dropping out from under you. If you put the life raft in the water and it's not tied to the ship, it's going to blow away. So that leaves you with one option: entering the water in a Gumby suit.

"But there isn't a surface to the water in those kind of conditions. Your ability to protect an airway so you can breathe is marginal at best with the sea foam, the spray, the rain water. I know rescue swimmers that have deployed in storms and they talk about how, within a couple of feet of the surface of the water, you really can't effectively breathe. And they've got a snorkel.

"So I thought it plausible that the person in the survival suit we found had been on the bridge when the ship went down. And everybody else who wasn't actively engaged in trying to save the ship had mustered in another compartment and was making preparations to abandon ship if it came to that. And then, I can't speculate, you know, beyond that."

Could the man in the survival suit have been Davidson? As the captain of the ship, he might have been the only person on the bridge when the ship was going down.

By Wednesday, October 7, a combination of air and water assets had searched a total of 274 hours, covering 195,601 nautical miles. They recovered an empty *El Faro* survival suit, a pair of soft-shell coolers, a reefer cargo door, a Minnie Mouse doll along with dozens of other toys, three life rings, one blue plastic parking curb, and an adult-size life vest.

A Carnival cruise ship sailing through the area encountered the debris field; passengers stood in silence on the decks as their ship cut a swath through the remains of *El Faro*—a miles-long oil slick.

CHAPTER 23

PROFIT AND LOSS

The US Coast Guard wasn't always synonymous with safety. It took a massive movement to create laws that offered protections for the men and women working on America's ships.

Wildly lucrative, economically necessary, and extremely dangerous, eighteenth- and nineteenth-century shipping occasionally ended in disaster, but the lives of sailors was not the young government's concern. Under federal law in 1790, for example, sailors could be imprisoned for deserting a ship before the voyage ended, even if they had very good reasons to flee. That law stood fast for more than a century. Indeed, flogging aboard ships was prohibited by law only after 1850; corporal punishment was abolished in 1897.

A merchant could load his creaky wooden vessel until its gunwales lapped at the brine, yet no one refused him passage. Along the docks, sober sailors eyed the teetering ship warily. Plied with drink or plagued by creditors, however, many men preferred to cast their lot with the open waters rather than submit to life in prison, either behind bars or behind a desk.

If a ship sunk somewhere out there on the seas or sharp bluffs, if it stubbornly refused to appear at port at its expected time—its crew and cargo buried in a watery grave—insurance would reimburse the merchant for his troubles. Even in 1835, Americans' boundless appetite for speed (and profit) did not go unnoticed. French diplomat Alexis de Tocqueville observed:

> "The European navigator is prudent when venturing out to sea; he only does so when the weather is suitable . . . [The American] sets sail while the storm is still rumbling; by night as well as by day he spreads full sails to the wind; he repairs storm damage as he goes; and when at last he draws near the end of his voyage, he flies toward the coast as if he could already see the port.
>
> The American is often ship-wrecked, but no other sailor crosses the sea as fast."

Likewise, the US Coast Guard had an economic, not human, mission. Formed by Alexander Hamilton, it was established to enforce customs laws rather than protect the men and women aboard the ships. Its officers boarded vessels only to check manifests to ensure that the government collected duties it was owed; if they saw a man float by clinging to a mast, they had no obligation to fish him out.

With the advent of steam, a series of catastrophic boiler explosions within full view of women and children ashore spurred the industry to do something about its iron-bellied beasts. In a single year—1832—14 percent of American steam vessels, most on the Mississippi, were blown to bits by their own shoddily built boilers. Following the Civil War, a ship burst apart, killing all thirteen hundred veterans aboard. The merciless sea had once been the greatest threat to the pursuit of happiness; now manmade machines consumed people and ships with nearly as much fervor.

But dangerous ships increased everyone's risk—financial risk, that is. Ship hulls were insured through industry-only clubs, such as Lloyd's of London. Whenever there was a loss, all shippers got hit with higher premiums. Someone had to wrestle this risk under control. So shipping companies decided to regulate themselves. The American Bureau of Shipping (ABS), a nonprofit founded in New York in 1862, was founded by merchants to establish design and construction standards for America's seagoing vessels. Like its British cousins, the society eventually adopted the Plimsoll mark; published wood-, iron-, and steel-hull guidelines for naval architects; and provided inspection services to their members in an effort to ensure America's merchant fleet adhered to high standards.

Throughout the nineteenth century, shipping was booming; the US population almost doubled each decade—from four million in 1790 to sixty-three million in 1890; concomitant with this unprecedented growth came unprecedented demand for boats and barges to move raw materials, manufactured goods, and people. But as the population spread inward from its coasts, opportunities on land superseded jobs on water. There are incredible photos of San Francisco Bay during the Gold Rush showing the harbor literally choked with skeletons of rotting sailing ships abandoned by mariners who'd seek their fortunes in the hills. Sailors on the East Coast had signed onto the ships in droves, rounded Cape Horn, and came up the California coast, only to desert their vessels as soon as they reached port.

Desperate for crews to hoist the sails or feed coal into steamships' ravenous boilers, companies hired third-party agents to coerce sailors, by hook or by crook, into signing on. In major ports including New York, San Francisco, Boston, and Baltimore, shipping companies teamed up with "crimps"—boardinghouse owners who took advances on sailors' wages in lieu of payment to cover their lodging and other expenses, including liquor and prostitutes. Shipping companies paid sailors' wages directly to the crimps, creating a system of indentured servitude. The companies also offered bonuses, known as blood money, to agents who signed men up for voyages. This system rewarded criminal methods. Agents regularly disguised themselves as sailors, offering rounds of drinks to unsuspecting seamen who then awoke from a liquor-induced haze aboard a ship, their future earnings already spent. Sometimes agents shanghaied sailors by force—clobbering them on the head, forging signatures, and dragging them unconscious onto a ship while the captain conveniently looked the other way.

Mariners fought these nefarious practices for years but had little sway over a corrupt political system committed to supporting the moneyed end of the industry. By the end of the nineteenth century, however, crusading mariners, especially foreign-born seamen, began to organize. Joining miners, factory workers, and other workingmen and -women, they fought for safer working conditions, better pay, and better protections. Their efforts were hampered by historically entrenched racism, especially on the East Coast where white men refused to join together with their fellow black sailors, but organizers eventually gained traction on the West Coast where most seamen were white Americans or European immigrants.

This movement—paid for in blood, sweat, and tears—laid the foundations for America's most resilient unions. The American Maritime Officers (AMO) union, the International Organization of Masters, Mates & Pilots (MM&P), the Seafarers International Union (SIU), and a handful of others continue to wield impressive power over the shipping industry a century later.

The sinking of the *Titanic* coupled with other major maritime tragedies in the early twentieth century, along with the successes of the organized labor movement, reinforced the need for a strong maritime regulatory agency. The volatile steam engine and its failings finally bound capitalism to workers' welfare.

From the moment labor organized, America's power brokers have hacked away at it. William Randolph Hearst, the Rupert Murdoch of his day, used his tabloid papers to disparage all unions, but especially the maritime unions. He called organized labor anti-American at best, communism at worst. Case in point: conservatives loathed the fact that merchant seamen received medical care through a system of public health hospitals first established in 1789. Seamen's hospitals were founded to prevent the spread of disease and, over time, grew to serve military dependents, coast guard personnel, and the poor. These hospitals were once financed by a tax on imported commodities, but as of 1884 that money went straight into the Treasury's general fund rather than specifically supporting seamen's health. After World War II, Republicans went to great lengths to dismantle the hospital system in support of their tax-cutting agenda. As president, Ronald Reagan finally succeeded in eliminating the hospitals from the federal books, and now the entire population those facilities once served, including America's mariners, receive care at private hospitals at private hospital rates.

Labor is a tiny fraction of shipping costs—about 4 percent. Officers and seamen are paid according to strict union rules. They receive regular training through their union halls and are represented by union lawyers when there's a dispute. They get their insurance and pensions through their unions, as well, and count on the unions to represent them in federal and industry negotiations.

Fuel is a major drain on shipping profits. To power a medium-size ship across the Atlantic demands as much as $40,000 a day. Every penny counts.

Unable to cut salaries enough to please their shareholders, shipping companies have saved a few dollars by cutting staff. Watches on the bridge were once divided among four mates; now three people share the twenty-four-hour watch period, meaning longer shifts and increased risk of exhaustion. In the days of Morse code, a dedicated radioman handled all shipboard communication. That position is gone; now the on-duty mate must man emergency messaging, along with all of his or her other responsibilities. That's how two full-time jobs magically become one.

Respected ship captains once retired ashore to become port captains—seasoned experts on land who advised captains at sea. They were well versed in the ways of the merchant marine, the weather, the ships, and the subtleties of staffing a vessel. Older mariners who worked with port captains say that in an apprenticeship industry like shipping, port captains' experience was invaluable in running a vessel. They'd discuss weather and routing and use their prodigious experience to guide younger captains in all things, including managing their diverse crews.

The man who filled that role for TOTE's ships, Bill Weisenborn, quit the company in 2012. He didn't want to relocate to Jacksonville and didn't like all this restructuring and downsizing. The port captain is a luxury of the past. Now port engineers— those trained on running the ship's engines—are expected to do it all. If an engine fails, it's catastrophic and impossible to ignore. An incremental breakdown of knowledge and leadership can go undetected for years.

As for regulation of a dangerous yet vital industry, all eyes eventually turned to the fifth branch of the armed forces, the coast guard. Over its two-hundred-plus-year history, the US Coast Guard fell under the auspices of various departments—the Treasury, the Department of the Navy, the Department of Commerce, the Department of Transportation, and most recently, the Department of Homeland Security.

Conflicting masters have led to conflicting allegiances. Since 1942 the US Coast Guard has been responsible for creating and enforcing laws to ensure that America's fleet—fishing vessels, ferries, pleasure boats, tankers, and container ships—remains safe and that its mariners are trained to deal with shipboard emergencies. But under George W. Bush following 9/11, the coast guard shifted focus yet again from vessel inspection and lifesaving to defense. Armed to the teeth, prepped for battle against innumerable enemies, the US Coast Guard now fights America's eternal war against drugs, terrorism, and illegal immigrants on the wide open seas.

When the coast guard became America's waterborne policing unit, spending much of its energy on drug interdiction rather than safety, the American Bureau of Shipping stepped in to take over the job of monitoring the commercial fleet. As a result of its growing responsibilities, ABS has grown exponentially to become a rich, politically powerful entity.

SITTING IN HIS WASHINGTON, DC, OFFICE ON OCTOBER 1, 2015, CAPTAIN Jason Neubauer of the US Coast Guard thought his service had much to be proud of. As chief of the Office of Investigations and Casualty Analysis, Neubauer was charged with probing major marine accidents to unearth their root causes, then suggest changes, if any, to the coast guard's myriad regulations in an effort to prevent future casualties. He was one part of the mechanism that ensures the safety of America's fleet.

Teutonic by way of California—tall and intimidating from a distance in his sharp navy blue uniform; at the other end of a handshake, warm and empathetic—Neubauer had eschewed his father's navy for the coast guard. His family gave him a hard time for joining the "puddle pirates," the derogatory term the navy uses for its coast-hugging cousins. But as Neubauer climbed the ranks to reach captain, a notch below rear admiral, at the age of forty-three, the teasing stopped.

The US Coast Guard's DC headquarters maintains a thick file of marine accidents, more than fifty-four hundred each year of varying magnitudes—from a tugboat engine explosion that killed the chief engineer aboard; to a giant oil drilling rig that broke free from its towlines in thirty-five-foot seas, beached on an island in the Gulf of Alaska, and threatened to dump 150,000 gallons of diesel onto the coastline; to the sinking of the three-masted HMS *Bounty*, caught in Hurricane Sandy, that took the life of Robin Walbridge, the ship's captain, and a crew member.

Among the cases in the coast guard's file, Neubauer knew of just three major American ship casualties in peacetime; none of those losses involved cargo ships like *El Faro*. One was the SS *Edmund Fitzgerald*, which sunk in a storm on Lake Superior in 1975. The second was the SS *Poet*, a bulk carrier which mysteriously vanished in the Atlantic in 1980.

The third casualty was the SS *Marine Electric*—a World War II–era bulk carrier loaded with coal that went down in 1983 when she sailed through a storm off the coast of Virginia, killing thirty-one of the thirty-four crew aboard. The ensuing Marine Board investigation was a massive undertaking and the final report detailed serious flaws in the ship inspection process.

The Marine Board found that the *Marine Electric*'s hatches and decks were so profoundly compromised by age, bad repair jobs, and neglect that they proved helpless against the Atlantic storm's pounding waves, yet the shipping company had faked inspection paperwork year after year to keep her sailing.

That 154-page Marine Board report became a seminal work, setting shipping regulations and standards for the next thirty-two years. Among industry changes that emerged from the report, the coast guard required that all ships begin carrying immersion suits for its crew when sailing in the North Atlantic during the winter. The coast guard's rescue swimmer program also emerged from the tragedy.

Perhaps most important, the *Marine Electric* report raised serious questions about the ABS's role in ship inspections: "Basically, ABS surveys and visits are oriented toward protecting the best interest of marine insurance underwriters, and not for the enforcement of Federal safety statutes and regulations," it said. "Since the cost of these surveys and visits is borne by the owners, or other interested parties, the attending surveyor is subject to the influence of such persons."

Ultimately, the Marine Board report warned the coast guard not to cede something as critical as ship inspection work to third parties. "For the purpose of enforcement of Federal Statutes and Regulations, [inspections] should be conducted by an impartial governmental agency having expertise in that field, with no other interests and/or obligations other than assuring compliance with applicable requirements. By virtue of its relationship to the vessel owners, the ABS cannot be considered impartial."

The report also questioned the depth of training US Coast Guard inspectors had received and recommended that the organization devote more resources to its inspections program. Mike Odom's career path from coast guard rescue swimmer to traveling inspector was directly influenced by the aftermath of the *Marine Electric* case.

Another fact emerged from the tragedy: the officers and crew aboard the *Marine Electric* had been well aware of the ship's horrible conditions but feared for their jobs "due to the lack of seagoing employment . . . they were content to sail the vessel." Even as far back as 1980, lack of job security in the US Merchant Marine colored mariners' judgment.

The *Marine Electric* investigation led to the scrapping of more than seventy elderly US-flagged vessels, and a much more robust system of checks and balances to ensure America's aging fleet remained seaworthy.

Sitting in his office that October morning, Neubauer firmly believed that better technology and rigorous inspections programs had made American shipping safer. He felt sure that the industry had greatly benefited from advances in weather forecasting, bolstered by satellite photography, dropsondes, and supercomputer-based modeling; satellite communication and GPS; and AIS tracking to precisely locate vessels. The coast guard's marine safety program had set strict deep-draft vessel standards, most of which had been entrusted to ABS. All this was supported by the best maritime rescue force on the planet—the coast guard, assisted by the navy and the air force—powered by the world's most sophisticated helicopters, planes, and cutters.

But at 10:00 that morning, Neubauer saw a notice about a search-and-rescue operation east of the Bahamas flash across his computer screen. He'd been tracking Hurricane Joaquin for the better part of a week and wasn't surprised that a vessel had been caught up in its clutches. He called District 7 to find out more. That's when he learned that the officers were trying in vain to reestablish communication with a 790-foot American vessel with thirty-three crew members aboard.

The fact that it was a large American container ship rattled Neubauer to his core. US-flagged ships may ground or collide with things, but they don't sail into hurricanes and vanish.

Was it possible, he privately wondered, that *El Faro* and her crew had slipped through the cracks?

First, he called the National Command Center to get the most up-to-date info. They didn't consider the ship lost, they told him, not yet, and were optimistic that it would pull through. Maybe the vessel's antennas broke off in the storm, they reminded him. That happens.

Then the captain walked down the hall to consult with Rear Admiral Paul Thomas. If *El Faro* sank, it would trigger a massive investigation, and he wanted his superiors to brace themselves for what might ensue. He and the admiral discussed the magnitude of a case like this—it would take years to gather evidence, interview all parties, analyze the data, and produce a report. They might uncover malfeasance, or the investigation could result in a complete overhaul of the coast guard's inspection program. It could be a generation-defining event.

Following protocol, Neubauer also called the head of Marine Safety at the National Transportation Safety Board located across the Anacostia River in central DC. The NTSB's marine safety division often worked in parallel with the coast guard to untangle the cause of accidents. That agency would produce a separate report and its own series of recommendations.

And then, like so many others—the families of those sailing on El Faro, the people working for TOTE, the men and women in District 7, the reporters deployed to Jacksonville, and mariners around the world—Neubauer waited, hoping for the best, preparing for the worst.

By end of day on October 2, they'd found no survivors and no ship.

In search of someone to lead the investigation, Neubauer began calling other coast guard captains whose experience made them good candidates. No one with the right qualifications could commit the time to something this big.

On the fifth of October, just before eleven o'clock in the morning, the US Coast Guard officially designated El Faro a major marine casualty. The ship was lost, no survivors. Thus began a multimillion-dollar effort to find out what happened to the vessel and the thirty-three people who had vanished with it.

Compelled by the mystery of El Faro and personally offended that a tragedy of this magnitude could happen in the modern era, Neubauer volunteered himself to lead the Marine Board's investigation. The admiral agreed.

CHAPTER 24

THE TRUTH IS OUT THERE

23.23°N, -73.55°W

In a small office at the NTSB in Washington, DC, sat Captain Neubauer's physical antipode. Compact, hair thinning, with an indeterminate midwestern accent, Tom Roth-Roffy would have blended right in to Washington, DC, circa 1955. The only thing missing from his otherwise classic government employee look was a pair of Robert McNamara glasses.

Tom was the quintessential hardworking, incorruptible federal investigator. He was raised in Panama, where his father was a navy officer, which no doubt fostered in young Tom a complicated relationship with authority. Cut off from mainstream American culture, he grew up in a republic created by an imperial-minded United States for the sole purpose of digging, and then overseeing, the canal that would ultimately unite America's coasts.

Tom's upbringing didn't make him cynical. It made him a firm believer in truth. His role model might have been the tireless Joe Friday, someone who holds undying faith in the meritocracy. And the indisputable nature of facts.

Tom was trained as a marine engineer and once served on ships laying cable around Puerto Rico, but most of his decades-long career has been spent solving complex mysteries for the US government. Surrounded by traces of the dead captured in black boxes, twisted hulls, oil spills, and mountains of data, Tom and his team track the human errors and engineering mistakes that lead to devastating losses. When describing his life, he uses headline cases—Deepwater Horizon or Exxon Valdez—to mark time the way most people use births, deaths, weddings, and anniversaries.

Tom was given the job of leading the NTSB's investigation of the *El Faro* mystery.

Five days after the disappearance of the ship, Tom joined Neubauer aboard a plane from DC to Jacksonville. The US Coast Guard and the NTSB would be working together to gather information about *El Faro*, but they would issue separate reports.

Many on the respective teams—naval architects, engineers, former mariners, marine inspectors—were on that plane, too. They'd worked together on previous cases, creating an instant rapport now. Both groups set up shop at the Jacksonville Marriott and spent their evenings huddled in conference rooms, comparing notes. They didn't want to burden the dozens of witnesses with more than one interview, or double efforts, so agreed to conduct most interviews jointly and coordinate information gathering.

El Faro's crew's wives, parents, and children who had come down during the tense weeklong search were also staying at the hotel when the investigative teams arrived. There were awkward moments when the investigators crossed paths with those dealing with shock and loss. Emotions were raw, and the proximity to family members made Neubauer even more committed to his work, he says.

Two days later, Rear Admiral Scott Buschman, commander, Coast Guard 7th District, announced the official end of the search-and-rescue effort: "I have come to a very difficult decision to suspend the search for the crew of *El Faro* at sunset tonight. My deepest condolences go to the families, loved ones, and friends of *El Faro* crew. US Coast Guard, US Navy, US Air Force, and the TOTE Maritime tug crews searched day and night, sometimes in perilous conditions with the hope of finding survivors in this tragic loss."

Some of the families were furious that the search was over. Others understood that all that could be done had been done.

Strong and resilient, like so many parents of New England mariners before them, Deb Roberts and Laurie Bobillot were among the first relatives of the crew to accept that *El Faro* had been lost. During the agonizing week in Jacksonville while the coast guard combed the ocean for the survivors—and then any evidence of the gigantic ship—they became the public face of grief.

They worked with the Red Cross to set up a Facebook page with regular updates for anyone connected to *El Faro*. They gave TV interviews to ensure that the human side of the tragedy wasn't lost in its telling. Pictures of Michael and Danielle—smiling in their crisp Maine Maritime Academy uniforms, Danielle's hair tucked beneath her hat—appeared everywhere.

Nothing captivates the American imagination like a shipwreck. Even after we lost our direct connection to the ocean, sea stories remain high in American lore. We still feast on the *Titanic* and continue to poke around its ancient remains. We consume the *Deadliest Catch*, a show shamelessly dedicated to tragedy at sea. Every year or so a blockbuster film depicts courageous sailors battling enormous waves. The ocean is a mystery, its forces unknowable. It draws us in.

Once the harrowing search for *El Faro* began, it dominated the news cycle. The media descended on the scene—vans from local news outlets, the AP, and national news outlets were parked outside the Marriott. Joaquin was the big story that week. The storm had caused historic flooding in South Carolina. Most Americans couldn't tell you what the merchant marine was, but during that week, the names of the seamen aboard the ship, and snippets of their lives and the people they'd left behind, continually ran online, on TV, and in newspapers across the country.

The incident instantly became politicized. Just a few days after *El Faro* was lost, Senator Bill Nelson of Florida called a meeting with Tom Roth-Roffy and Jason Neubauer at *El Yunque*, docked in Jacksonville. The senator demanded answers. *Why couldn't they find* El Faro? he asked. His questions seemed reasonable. *How*, he wanted to know, *in the twenty-first century, when every idiot has a smart phone with GPS, can we lose a huge cargo ship?*

Tom was accustomed to working with government operatives, investigators, and other officials, but not politicians. He told me that in his long career, he'd never witnessed a show of power like that.

Tom's earnestness seems anachronistic in this era of political bombast, but it would make him the perfect foil for the corporate tripe spewed by TOTE's executives and lawyers. In the case of *El Faro*, this small, unassuming man became something of a folk hero. That would ultimately lead to his undoing.

The NTSB and coast guard began compiling lists of people they wanted to talk to and documents they wanted to see. Without survivors, the lists were extensive; the inquiry was broad. They set up a conference room for interviews and began bringing people in—TOTE executives and employees, family members, former and current TOTE mariners, the coast guard's search-and-rescue teams, pilots, marine inspectors, and naval architects. One by one, they told their tiny piece of the story. A story began to form about an outdated vessel and a shipping company in transition, but none of that would in itself sink a ship.

The most compelling leads were the brief conversations Davidson had had with Captain Lawrence's voice mail and the emergency operator. Something about a blown scuttle, a list, lost propulsion, dewatering. His words were put under a microscope, analyzed and reanalyzed, but that inquisition only led to more questions. Why was the ship there in the first place? And why did she sink?

It was clear: the investigators needed to get *El Faro*'s black box—the voyage data recorder, or VDR for short. The VDR had been bolted to the roof of *El Faro*'s wheelhouse, at the highest point of the ship. The newest VDRs are designed to break free and float in the event of an accident, but this one wasn't.

That meant that the VDR was somewhere deep in the ocean off the coast of the Bahamas. The coast guard had an approximate location based on the ship's last known position, the debris field, and the navy's hydrophones, but that didn't mean they could pinpoint it on a map. It was somewhere in a two-hundred-square-mile area, about forty miles east of San Salvador.

If they were going to find the ship, the investigators had to move quickly. *El Faro*'s VDR had been equipped with a pinger—a weak battery-powered beacon with a life span of about thirty days after activation. They had to get to the beacon before the battery died. Otherwise, finding the VDR could turn into a huge ordeal.

The NTSB had a member agreement with the navy to request help in locating wreckage or data recorders. The navy, for example, assisted in finding aircraft wreckage after TWA flight 800, a Boeing 747, exploded shortly after takeoff from New York's JFK airport in 1996.

Search money was quickly approved by Congress with major support from Senator Jay Rockefeller, a champion of the American shipping industry.

On October 19, USNS *Apache*—a Powhatan class of tug—deployed out of Little Creek, Virginia, equipped with a suite of underwater detection equipment and a team of technical experts to run it. The crew's mission: search nearly two hundred square miles of the sandy ocean bottom fifteen thousand feet below for a cargo ship.

Tom Roth-Roffy was on that mission. He boarded the 226-foot vessel in Virginia and rode it nearly one thousand miles down to the warm waters off the Bahamas. Once the boat arrived in the search area three days later, the technical team aboard the *Apache* towed a seventy-pound, bright yellow stingray-shaped pinger locator back and forth along the search pattern, hoping to pick up a signal from *El Faro*'s beacon.

The nerve center of the operation was set up in a narrow, windowless room in the *Apache*, where an array of outboard gear and servers stacked in racks along one wall fed nearly a dozen computer screens of various sizes.

The pinger locator had to be within a certain distance of the beacon, so the technicians towed it far below the *Apache* on a long cable. Though it was sensitive, the locator couldn't be rushed. The initial two-hundred-square-mile search went at about two miles per hour. When the *Apache* reached the edge of its search field, the boat above had to reverse direction, which meant waiting for the locator to swing back and settle into position before continuing. After four days of dragging the locator twenty-four hours a day, the technical crew aboard the *Apache* heard nothing from *El Faro*. The massive container ship had seemingly vanished.

The *Apache* team was confident that they were close enough to *El Faro* to pick up the pinger's signal if it was working. Clearly, something was wrong.

They switched to using a side-scan sonar called Orion, a thirty-six-hundred-pound, truck-size machine that sends pulses down into the deep to produce contour images of the seafloor on the technicians' computers. Orion had been strapped to the *Apache*'s deck, and the crew used the boat's crane to carefully lift it up and over the side, then gently release it into the sea, its fiber-optic umbilical cord trailing behind it as it traveled down into the void.

Again, with every direction change following the search field, they had to winch Orion almost to the surface, turn the boat, then send Orion back down and wait for it to stabilize before proceeding. It was tedious work, projected to take fourteen days to complete.

The *Apache*'s crew was divided into six-hour watches to monitor the computer screens for anything that looked unusual.

After five days of dragging, at 1:35 p.m. on October 31, a technician saw something new: right angles on the otherwise indeterminate seafloor. Then he saw shadows. He jacked up the resolution of Orion's imagery to reveal what looked like a tiny object the size and shape of the missing container ship. It was unmistakable.

Tom got the wake-up call as soon as the vessel was discovered. He hurried over to the control center to watch the recording of the initial sonar echoes. On the computer screen, Tom clearly saw a container ship resting upright on the ocean floor. He could make out what appeared to be the ship's house and the exhaust stack. Finer resolution revealed that *El Faro* had landed with such force that the sandy ocean floor around it was frozen in the shape of a permanent splash.

El Faro and her secrets lay directly below him.

Tom felt a great sense of relief wash over him. The investigation, one of the most important of his career, now had an accident site that he could study. They could scan the hull to determine whether it had been breached. They could examine patterns on the seafloor to get a better understanding of how it went down. Even better, the ship's VDR might be accessible somewhere near the wreck. That black box was the holy grail of any investigation. Tom was one step closer to solving a case. He was elated.

After locating the hull, the techs slowly winched up Orion, and then sent down a CURV-21 to document the site. The eight-foot-high, sixty-five-hundred-pound, fiber-optic cable-controlled vehicle was capable of withstanding three tons per square inch, the pressure at *El Faro*, three thousand feet deeper than the *Titanic*. The force down there was so great that it caused the heavy steel containers that were once lashed to the ship's decks to crush inward as they sunk.

A primitive version of the CURV-21 had been used to help the NTSB learn more about another famous shipwreck, the SS *Edmund Fitzgerald*, a 729-foot freighter that disappeared in 1975 while sailing in a severe storm on Lake Superior. All twenty-nine crewmen aboard perished; like *El Faro*, there were no eyewitnesses.

The great tragedy of the *Fitzgerald*—the worst peacetime accident on the Great Lakes—led to major industry changes. It was a benchmark case for the Marine Board of Investigation, one that Captain Neubauer and Tom Roth-Roffy thought a lot about when *El Faro* disappeared.

Ultimately, the cause of the *Fitzgerald*'s sinking was inconclusive, and the event continues to draw armchair theorists, but many of the contributing factors debated during that case came up again in the *El Faro* investigation.

It's possible that opened or damaged hatch covers may have caused flooding in the thirty-five-foot waves that ultimately overwhelmed the ship, the 1970s US Coast Guard report theorized. This flooding could have been slowed by the installation of watertight doors between the vessel's holds, but the shipping industry had fought against such protective measures for years, claiming that such modifications would be prohibitively expensive. Without watertight separation between her holds, the *Fitzgerald* probably went down very fast. The captain didn't even have time to send a distress signal.

The *Fitzgerald* was also overloaded—her Plimsoll line had been raised three times in a few short years leading up to the accident, allowing her freeboard to run three feet closer to the water than specified by her original designers. That made her very vulnerable to flooding in foul weather.

Another problem revealed in the case: the National Weather

Service had inaccurately forecasted the conditions on the Lakes during the storm that sunk her. In fact, the seas were much higher than predicted and may have caused a few massive rogue waves.

There were human factors involved as well. The *Fitzgerald's* captain was known for pushing his vessel hard and flouting storm warnings. When weather kicked up, he'd thrill to it. He boasted about the strength of his ship and the bravado of his crew—they could take on anything Mother Nature could throw at them. In his book *Deep Survival*, Laurence Gonzales writes, "The word 'experienced' often refers to someone who has gotten away with doing the wrong thing more frequently than you have." This captain was experienced in exactly in that way.

The sinking of the *Fitzgerald* showed the coast guard that shipping companies weren't capable of regulating themselves; the organization needed to step up to protect the men and women who crewed these vessels. Among the many new regulations implemented following the sinking: shippers had to provide survival suits for all hands sailing on the Great Lakes to stave off hypothermia and drowning; Great Lakes navigation charts were improved; vessels above a certain tonnage had to be equipped with depth finders and EPIRBs; and the coast guard committed to boarding and inspecting all Great Lakes ships for watertight integrity each year before the fall storms kicked in.

These new regulations served the industry well for decades after the *Fitzgerald* catastrophe, with the exception of the sinking of the *Marine Electric* in 1983. American commercial shipping enjoyed a nearly spotless safety record, with no major losses after that.

But almost forty years to the day, the specter of the *Fitzgerald* returned when *El Faro* went down. Had regulators and industry slacked off in the ensuing time? Had we all become too complacent?

That's how Tom Roth-Roffy found himself in the dark, chilly

control room on the *Apache*, transfixed by the images being sent back to his computer screen from the deepest Atlantic by the CURV-21.

The CURV ran tethered to the *Apache* by a thirty-six-thousand-foot-long electro-optical-magnetic cable that wrapped around an enormous winch on the vessel's deck. Using the CURV's thrusters, the techs could fly it like a drone, but very slowly.

When the CURV was in position close to the sunken hull's location, the techs switched on its high-resolution cameras and bright lights to illuminate the abyss. A murky gray-blue landscape appeared on their screens. If the CURV got too close to the sea bottom, its thrusters kicked up fine particles that got caught in its bright spotlight, temporarily blinding them.

The first thing the CURV revealed was the miles-long and -wide path of debris and containers. As the loaded freighter went down, everything she held peeled off, leaving a trail for the CURV to follow, like breadcrumbs in the forest. Some containers opened as they made their rapid journey three miles to the bottom, scattering their contents onto the soft ocean floor. Cars spilled out of the ship's hull in slow motion, their components breaking free and raining down. The site looked like a murky gray sprawling city made entirely of junk.

Tom saw a bicycle. Car batteries. Crumpled containers everywhere. Flung about was the eerily familiar detritus of modern life—a printer, a microwave, the top of a car, abstractly rendered in the foreign dusty desert of the deep Atlantic.

Eventually, they came upon the hull—a steep wall of solid steel.

In the quiet calm of the ocean's depths, the CURV glided along the port side of the ship, recording the frozen remains of a calamitous event. Then the CURV reached the stern. It turned the corner and slowly panned up. Emerging from the blue gloom, white letters stood proud on the twisted hull: *EL ARO, San Juan PR*. The steel where the *F* had been was crushed inward.

Tom and the tech team gazed at those words in silence.

After lingering for a few minutes, the CURV moved up to the vast top deck, now empty of the cargo boxes that once were stacked three-high. It ran down the length of the deck like a plane taxiing on a runway until it reached the multistory wheelhouse where the ship's crew once lived and worked. The CURV's operator pivoted its camera and lights upward and began puttering up to the top of the house, where the VDR was expected to be. As the vehicle rose, those watching aboard the *Apache* shuddered.

El Faro's top two stories, including the navigation bridge, mast, and VDR, were gone.

From the way she'd landed, the team theorized that after *El Faro* sank below the surface, she spiraled down at about 45 miles per hour stern-first at a steep angle to her final resting place on the ocean floor. She descended so quickly that at some point, the navigation bridge must have torn away. Later, another theory would be put forth: that when the cooler waters hit the ship's boilers' superheaters, there was an explosion that blew off the house.

After weeks of searching, the NTSB had found the ship, an important first step. But Tom's mood was solemn. The VDR was attached to the roof of the navigation bridge, and that piece of the ship was nowhere to be seen. They hadn't located the one thing that would unlock his investigation. If they couldn't find the VDR, the NTSB and the coast guard would have to rely on conjecture. But how would they find something the size and shape of a coffee can in miles and miles of wreckage?

CHAPTER 25

HOW TO SINK A SHIP

When *El Faro* disappeared, John Glanfield, the retired shipbuilder, then seventy-five years old, pulled out the stack of ancient Sun Ship blueprints that he kept in his basement. He used the drawings to begin looking for flaws in the ship's design that could have sunk her. Driven by curiosity and anguish, he needed to know. He needed to understand how one of his beloved children could end up three miles down, taking thirty-three people with her.

John sat at his kitchen table making a paper model of the ship's midsection, trying to reconstruct what might have happened to her on the morning of October 1, 2015. Building models came naturally to John. Throughout his life, he'd nurtured his artistic tendencies by drawing cartoons on Sun Ship drafting paper and crafting models to amuse his shipyard colleagues.

When his kids were little, John would put them to bed and disappear to the basement, working for hours on tiny versions of the ships he'd built or admired. He once created a five-foot-long *Titanic* for a young man who believed he was the reincarnation of someone who'd died when the great ship went down. When John went to deliver the model to his client's Philadelphia apartment, he was amazed to see that the rooms had been done just like a cabin of the ill-fated passenger ship, complete with heavy drapes, brass, and oriental rugs.

John developed a yen for crafting ships in a bottle, too. He made a model of the *Hughes Glomar Explorer*, the vessel that recovered the Soviet sub. His model was built into the oversize lightbulb that once sat atop Sun Ship's tallest crane to warn away aircraft. To complete the model, John had to imagine what the mysterious CIA-commissioned claw looked like because he'd never seen it; nor had anyone he knew. The miniature "drilling rig" rests atop a sliver of blue sea. Below it, a thin cable hoists a yellow five-armed claw; in its clutches is a black-painted sub, carved out of wood.

When John heard about the sinking of *El Faro*, he built several *El Faro*s in a bottle for families of the deceased. Red, blue, and gray hand-lettered "Sea Star" boxes are stacked on the deck. A tiny plume of cotton trails from the ship's red and blue stack.

John had a hard time getting addresses of the families to send them his ships. So he turned to Facebook where he found Deb Roberts, who'd lost her son, Michael Holland. Deb helped connect John with others, and the little models went out across the country—Maine, Florida, Massachusetts. One went to Arizona to Captain Earl Loftfield, master of *El Yunque*.

John developed a theory about *El Faro*'s fate and wanted to share his thoughts with me after reading my article about *El Faro* published in *Yankee* magazine in October 2016.

When we first spoke on the phone, John kept referencing something called the "downflooding angle." After hanging up, I looked up the term in the *Code of Federal Regulations*—a multi-volume compilation of all US rules covering every conceivable industry from education, to energy, to agriculture, to foreign relations, to shipping. The downflooding angle refers to how far you'd have to tip a boat in calm conditions for water to penetrate the boat's first nonweathertight opening. In a simple sailboat, that would be the moment when the boat tips enough that its gunwale dips below the waterline. Water then floods into the hull.

I once capsized a small boat this way. On a very windy day, I accidentally exceeded the boat's downflooding angle while executing a jibe. Wind caught my sail and pulled the mast toward the water. I'd always popped up before. But when the gunwale dipped below the water, the river came pouring in, pulling the boat into an unrecoverable roll. In an instant, I found myself clinging to its side, looking for a rescue launch. Once it tipped, there was literally nothing I could do but slide into the river and wait for help.

Now that I understood the term, it seemed like such a critical part of sailing that I assumed every ship's officer would know his or her vessel's downflooding angle.

So what did this have to do with *El Faro*?

One early April afternoon, I visited John and his wife, Dot, at their immaculate suburban Philadelphia house. After lunch, we cleared the kitchen table to make way for John's forty-year-old blueprints. After years of working with power tools in resonant steel hulls, he can't hear that well; he spoke for three hours in a nearly unbroken stream, walking me through every aspect of the ship that he thought might have come into play that October morning.

He'd been thinking about this for a long time and was excited to share.

"When you look at *El Faro*, you see the big blisters on the side, forward of the house," he says. "There are three of them on each side. They stick out from the hull. And those were the ends of the ventilation shafts. I've never, ever seen that on another ship—air shafts external to the hull."

Inside those blisters—metal boxes hanging off the ship—were big fans designed to blow air down into the deepest part of the vessel, or allow air to escape out of the vessel. If you didn't ventilate the cargo areas, poisonous and flammable gases from the cars and trucks stowed below could build up, risking fire or an explosion.

John continued: "When you look at how the ship was designed, you say, *oh my gosh*. Because you see these big ducts through here?"—he points to some ovals on his blueprints, just below the deck line—"They go all the way to the bottom of the cargo hold. There are special rules that allow this vent to be [only] this far above the waterline and still maintain watertight integrity. And that's all calculated into what's called the downflooding angle."

In modern vessels, ventilation stacks rise up from the main deck. But the designers of the *Ponce*-class ships wanted to keep the decks as clear as possible. Otherwise, trucks and trailers could hit the stacks during loading and unloading. The designers were also working with less stringent codes.

So they hung the ventilation boxes off the side of the ship below the main deck line. Inside each of these boxes was a sizable duct that penetrated through the ship's thick steel hull, then turned straight down, running deep along the inside wall of the ship and into the cavernous holds.

It was like having your nostrils somewhere near your belly button.

John picks up the flimsy paper model of *El Faro*'s midsection that he made a few months after the ship disappeared. He built it so that he could study the vessel's inherent vulnerabilities, and it takes me a while to get oriented because it's missing context—no bow, no stern—more like a fragile box with some dollhouse-size ramps and a few big holes in the sides.

As John points out various things to me, the model flops around in his large, rough shipbuilder's hands. The paper floor of the second deck isn't attached to the paper hull, adding to my confusion. But in moments of clarity, I see the ship, and I see how a serious

list would position these unprotected openings dangerously close to the waterline. *El Faro* was an accident waiting to happen.

"So here we are with the ship that's all intact and everything is fine as long as the ship is sailing," John says. "But then all of a sudden, the downflooding angle comes into play. When? At what point in time?"

He rolls his model to the starboard side. And now I notice that he's drawn a blue line indicating where a calm sea would be on the ship at a 15-degree list. The line—marking the point of no return, just like on my sailboat—wraps around the starboard side. Then I feel my gut tighten. The waterline goes right over the ventilation openings in the hull at the second deck.

With a list like that, you couldn't stop water from getting in. It isn't just washing across the deck. It's pouring into the ventilation shafts straight down into the holds, flooding the ship. The vessel would never be able to right itself. Like the sailboat, it would just lie on its side, taking on water, until it drowned.

"And remember," John says, "the hold was one big open space." Unlike some modern ships, where the cargo holds may be divided into starboard and port compartments, *El Faro*'s holds were divided into sections along her length, but not her width. Any water that got into a hold would slosh uncontrollably from side to side, compounding the ship's movements. This is called the "free surface effect," and it can have a catastrophic impact on a vessel's stability if she's rolling and pitching in high seas.

The fact that *El Faro* was designed this way hadn't been a problem for forty years. Like many roll on/roll off ships of her era, she had lots of holes punched into her hull, all of which complied with the codes of the day. That said, regulations for this class of ship had changed over the ensuing time due to sometimes fatal accidents. Their weaknesses were well known and documented. But *El Faro*'s design had been grandfathered in.

"At what kind of point when the ship starts to go down by the bow does the downflooding angle come into play?" To answer his question, John tilts his model to show pitch—the bow rising up the crest of a wave, then diving down into a trough. That back-and-forth movement gives water further opportunity to exploit *El Faro*'s unprotected openings. Combine a pitch and a roll, and it doesn't take a lot to sink the ship.

To build his model and make his calculations, John assumed that the ventilation trunks were intact. The fan boxes hanging off the ship's side were built with a thirty-five-inch-high steel baffle that forced air to take an indirect route into and out of the ship. Air had to rise up and over the baffle, which, due to its height, gave *El Faro* a watertight envelope fourteen feet, eight inches above the waterline, as required by regulations. On paper, at least, she could survive a 15-degree list.

In reality, over the years that *El Faro* sailed, the function of those baffles had been lost. When it was found that sea spray was getting inside her fan boxes and rotting away the steel, someone decided to simply drill bigger holes at the base of the baffles to allow the water to drain. Problem solved.

The result was that when she sailed, *El Faro*'s first unprotected opening was three feet closer to the waterline than anyone thought. The downflooding angle was no longer 15 degrees. It might have been more like 10. What would it take to sink her? Not as much as it should. A few hours in big seas that got her pitching and rolling would definitely cause water to flood her holds. Enough water, and she'd no longer act like a balloon floating on the surface. She'd act like a weight.

CHAPTER 26

ADMIRAL GREENE CLEARS THE AIR

On Monday, February 16, 2016, the Prime Osborne Convention Center in Jacksonville became the epicenter of the *El Faro* drama.

The white neoclassical building with its stately colonnade, built in 1919, sits inexplicably alone in a vast, sunbaked emptiness.

The building is a haunting reminder of America's promise and prosperity, racism and decline. It had previously served as the bustling New Union Terminal Building—once the largest rail station in the South. Before the automobile took over the American landscape, trains at Union Station served as many as twenty thousand passengers a day traveling north to the great cities or south to the warm Florida coast. Wealthy New Yorkers and poor workingmen and -women briefly passed each other on their way to resorts in Miami or job prospects throughout the nation.

Where there was once an expanse of tracks carrying up to 142 trains a day is now cracked pavement. The last train pulled out of

Union Station in 1974. Like a thumb in the eye, Florida's Interstate 95, touted by the *Miami-Herald* in 1956 as a "slum-clearance project," opened in 1960 just a few hundred feet from the station, sounding the American railroad's death knell.

Nearby, the Jacksonville Skyway, a depressingly underused monorail built in 1986, clatters back and forth along its two and a half miles of track. The monorail was the one realization that emerged from a feverish urban vision once imagined for Jacksonville. Instead of blossoming into the Chicago of the South, however, downtown feels more like Detroit—rife with abandoned buildings and abandoned lots, scored by multilane roads occupied by indifferent drivers. Although the Skyway is free, it attracts few riders. At rush hour, my only companion was a homeless man.

The largest room in the Prime Osborne Convention Center once served as the main ticketing hall for the train station. With soaring vaulted ceilings and decorative friezes, illuminated by tall windows and large clerestories, it still feels like a grand railroad station. You can easily imagine the ladies and gentlemen in their finery eagerly scanning the timetables and purchasing tickets for their journeys across America.

There's no direct connection from the main hall to the second-largest room at the former station—you have to pass through a labyrinth of closetlike spaces to get from one to the other. Formally, one entered the smaller hall via a modest side entrance off the street. The smaller hall's ceiling is lower, coffered but flat. Its finishes are basic. It's not much more than a room.

This is where, under Jim Crow, Jacksonville's black passengers were permitted to buy their tickets and wait.

And this is where the Marine Board of Investigation *El Faro* hearings were held.

That first morning of the first round of hearings, family members, investigators, the media, and witnesses arrived before

9:00 a.m., parked in a gravelly lot in the shadow of the monorail, and made their way to the smaller hall.

Once inside, there was a deep sense of gravitas. Long vertical blinds blocked the Florida sunshine. A few dozen rows of stacking chairs faced a temporary podium framed by a navy blue velvet backdrop. The American flag and US Coast Guard flag stood side by side behind a long table outfitted with a pleated taupe skirt. Captain Neubauer sat in the center of the table flanked by coast guard investigators and members of the NTSB. The session opened with a moment of silence for *El Faro*'s thirty-three crew members who had perished with the ship.

Occupying the first three rows on the left half of the hall were the wives, parents, and children of the dead. They would camp out there all day, every day, for a total of six weeks. They constructed temporary shrines—propping up framed photographs of the men and women they'd lost—to remind TOTE's executives and lawyers, Captain Neubauer and the coast guard, and all parties of the human side of the loss. Lives were broken. They wore custom T-shirts emblazoned with EL FARO 33 remembrances. They listened respectfully, took notes, and occasionally shook their heads in disgust during testimony. Their loved ones were gone; they were there to find out why.

Most of the people occupying those first three rows were related to the unlicensed seamen, including Jill Jackson-d'Entremont and Glen Jackson who'd lost their brother Jack Jackson; Marlena Porter, wife of AB James Porter; Val Champa who'd lost her son Louis Champa; Rochelle Hamm, wife of Frank Hamm; Pastor Green, stepfather of LaShawn Rivera; and Gina Lightfoot, sister of the boatswain Roan Lightfoot. For the most part, they lived near Jacksonville. Up north, where most of the officers resided, those left behind tried their best to pick up the pieces, catching snippets of the hearings when they could. A few officers' relatives, including Rich Pusatere's father (retired and based in New York) and Steven Shultz's wife (who lived in Florida) attended many of the hearings.

The media sat on the other side of the room. TV cameras lined the back. Two court stenographers sat at the ready. In front of the proscenium was a witness table equipped with a couple of microphones.

Around the world, mariners watched the hearings via Livestream.

The first witness was Phil Morrell, vice president of Marine Operations, Commercial for TOTE Services. He was an executive high up on TOTE's organizational chart responsible for the operations of the company's fleet of ships. The line of questioning quickly established that Morrell had no direct marine experience. He was a budget guy, a numbers guy. He described a multitude of entities within his company—Saltchuk, TOTE Services, TOTE Maritime Puerto Rico, TOTE Maritime Alaska. He sometimes struggled to name his fellow executives' titles and responsibilities, evidence of the recent restructuring. Until 2013, he'd only been responsible for the Alaska trade. Following the firings and corporate shuffling, his workload doubled. He now had to oversee the company's Puerto Rican outfit from Tacoma, Washington, thirty-seven hundred miles away.

For the past few years, Morrell said, he'd been focused on getting the new LNG ships ready for Puerto Rico. He knew remarkably little about ship operations—how cargo got ordered and found its way to the ship, how it was loaded, and how it got to its destination. But he wanted to make one point very clear: the ship's captain was in charge of the vessel at sea. "The master is in charge of the voyage plan and voyage preparation," he said. If the captain had a question about anything while en route, he was basically on his own.

From the point of view of admiralty law, this was a smart position for Morrell to take. The Limitation of Liability Act, passed in 1851, caps a shipowner's liability to the value of the ship, if the accident is not caused by the vessel owner's neglect or malfeasance. In other words, as long as a shipping company doesn't interfere with its captain's decisions while he or she is at sea, explains admiralty and maritime lawyer Chris Hug, vessel owners can limit their liability when the causes of the accident occurred without their "privity or knowledge."

Shipping was a risky business throughout the nineteenth century; Congress believed that its role was to shelter owners from lawsuits and egregious payouts. Now much of American admiralty law focuses on who is responsible for what, and much of it favors the shipowners.

One could say that before the age of satellite communication, the Limitation of Liability Act made a certain amount of sense. When a vessel was out of sight of land, its owners had no means of contacting it. At that point, how could they prevent their officers from making fatal decisions? Holding a shipping company accountable didn't seem fair. But these days, the law seems profoundly anachronistic. It could even encourage deliberate negligence.

When at sea, TOTE's ships sent noon reports via email to nearly everyone in the office, but no one was specifically charged with monitoring these reports. With the elimination of the port captain—the one member of the organization who would have been fully dedicated to the ship's operations at sea—there was a gaping hole in management. When *El Faro* steamed into Joaquin, no one at TOTE was specifically responsible for watching the weather or tracking the ships, and from a liability standpoint, that fact absolved TOTE from responsibility.

After the sinking of *El Faro*, those who tried to make claims against the company—either for loss of life or loss of cargo—came up against the limitation law. Families of the dead were dismayed to discover that they had the burden of proving that the company's deliberate malfeasance led to the vessel's unseaworthiness, which led to its undoing. The law even precluded the plaintiffs' right to a jury trial. All families would eventually settle with the company for undisclosed amounts.

During the hearings, the coast guard repeatedly tried to find ways to hold TOTE responsible for the accident. Morrell was asked, "Who at TOTE oversaw things related to the ships' operation such as voyage planning, anchoring, maneuvering, pilot-master exchanges, and the route ships are taking at sea?"

"That's managed on board by the captain," Morrell answered. "That's all managed on board by the captain," he repeated. "He has total responsibility for all of that work." The men and women behind him rolled their eyes and squirmed in their seats. So this is how it was going to go.

Morrell, along with all of TOTE's executives, maintained that they followed the letter of the law. They adhered to all ship inspection guidelines and paid for any necessary repairs. Their paperwork was flawless. On board, their crew followed the international standards for watch-keeping. If anyone was overworked aboard *El Faro*, it was their own fault.

If Morrell's responses seemed slightly defensive, his boss's testimony seemed the epitome of executive evasion. Throughout Rear Admiral Philip Greene's full day of testimony, he continually praised the managers of his company in breathless tones, extolling their leadership qualities, praising their expertise and supreme focus on safety. As the day wore on, his platitudes must have been sickening to those seeking contrition from the company that had allowed not one—but two—of its ships to sail into a hurricane.

Tom Roth-Roffy spoke for many when he expressed confusion about the labyrinthine corporate structure.

Greene took the opportunity to clear the air: "In the fall of 2012, TOTE Services, as part of branding alignment, underwent a name change to TOTE Services. And in February of 2013, I was appointed president of TOTE Services, with a charge to grow the company. Because it's a phenomenal company with a great base of business, fine reputation, but an opportunity to grow as well within the field of vessel management.

"A part of that change of management involved a couple of things. One: assimilating technical management of the Totem and Sea Star line vessels under one entity, which happened to be TOTE Services, a vessel management company. Another initiative was reviewing and ultimately deciding to relocate the company from where it had been located within the New York, New Jersey, Philly area for almost forty years. In my humble opinion, Jacksonville was a positive place to come because of the workforce here, the synergy with a sister company that was here, and proximity on the East Coast. It was also a part of assimilating what was called TOTE—Totem Vessel Operations which provided an excellent opportunity to revisit the structure of TOTE Services's organizational structure.

"Along with that we also had a very important, in 2013 and into 2014, effort for financial backbone alignment with our shared service platform and basically, taking the ethos and leadership philosophy of our leadership team and spreading them effectively across our organization. So it was very exciting as we undertook these change management things, but it was done in a very methodical planned way, realizing first and foremost that safety programs and other key elements that sustain fleet readiness are kept uppermost in our minds.

"Obviously we had employees that chose to move with us from New Jersey to Jacksonville, but some didn't, which afforded us the opportunity to then bring on new and talented employees from the Jacksonville area. And we also brought carryover employees who had great backgrounds and experience from the Totem and Sea Star organizations as they assimilated into TOTE Services.

"In July 2014, we stood up our new headquarters here in Jacksonville in what I thought was a well-planned-out, seamless transition in a purpose-filled space that fit my philosophy of transparency, communication, and teamwork. And then, from that point [we] have continued to emphasize business processes, continuous improvement in all that we do, which by the way is a cornerstone of the ISM system. And we are now well established and very pleased to be here in Jacksonville. Great city and great people."

Tom Roth-Roffy wasn't sure what to do with all that corporate-speak, so he continued to pursue hard facts. He wanted to know why Captain Davidson's file was thin, and why he'd received only one official job evaluation during his two-year tenure with TOTE. If the company was working as well as Greene implied, things like captains' evaluations should happen like clockwork, especially when they're being considered for new positions.

Tom asked, "Is there some management process to track the completion of performance evaluations within the HR function?"

Greene replied that, as president of the company, he didn't expect such petty paperwork issues to rise to his level. Besides, he added, the officers aboard his ships had attended marine academies and were licensed by the US Coast Guard. Why would anyone ever question their judgment?

"I would like to think that the United States Coast Guard under Homeland Security delivers requirements that are respected, that are creditable, and that we're proud of . . ." Greene said.

"And I would hope that our credentialing process represents the fact that we're producing the finest officers to sail on our ships [that] the world can provide. And I know that our training institutions—from the United States Merchant Marine Academy, to our state academies, to the Coast Guard Academy, to our other institutions—we all could be proud that we deliver for our country. I believe that the American Maritime Officers and the Seafarers International Union [the unions that TOTE uses to staff its vessels] represent a gold standard. I'm proud of our officers and our men and women who sail on the ships."

Sitting in the hearing room, I found it fascinating to watch corporate America wrangling with government America. Members of the Marine Board were looking for answers in earnest; highly paid shipping executives were filibustering. Pricey lawyers representing TOTE were running interference. It was clear that Greene could go on all day about how proud he was of his company. He seemed perfectly comfortable spewing nonsense that flew in the face of facts.

Captain Neubauer pressed on, trying in vain to pinpoint specific issues within the company's policy that could have led to *El Faro*'s ill-fated voyage. "If the TOTE Services master communicated a voyage plan that was considered manifestly unsafe by the company," he asked, "who at the company has the duty and the authority to intervene? Who would have that authority?"

Greene replied, "With all due respect, Captain, our masters wouldn't do that."

The profound irony washed over the room. "[TOTE's captains] wouldn't create a voyage plan, in my estimation, that was anything but based on their immense credentials, their qualifications, and their recognition for the voyage at hand."

Had the voyage of *El Faro* sailed into a hurricane of fake news?

Then Greene passed the buck. It was as if this whole business was beneath him.

"Sir," Greene continued, "I'm the president of the company with twenty-six ships in a fleet. I provide strategic direction. I have qualified individuals within the portfolio areas, officers in the company. I expect them to dedicate the time that is required and necessary and appropriate to fulfill their duties in the capacities that they serve. And I rely and trust and count on them to do that, as I oversee a *significant* company."

Later that day, Commander Mike Odom, the former rescue

swimmer, focused on procedural operations once again. Odom was one of the coast guard's few traveling ship inspectors who go around the world boarding vessels to make sure they comply with the most current regulations. He'd seen a lot of questionable practices in the American merchant marine.

Odom asked whether TOTE had developed contingency plans that would guide how shoreside personnel dealt with a vessel in specific emergency situations. He was thinking about the hurricane—a very specific situation, but an expected one, considering the fact that TOTE sailed in the Caribbean. What was TOTE's directive to its employees when a massive hurricane threatened its assets?

Once again, Greene replied that TOTE followed the letter of the law, implying that if there was a failure, it wasn't the company's fault, it was the law's. He said in no uncertain terms that TOTE's safety management system (SMS) complied with the US Coast Guard's regulations. Which it did.

In fact, those regulations are pretty loose. Instead of providing a model or clear guidelines, the coast guard merely requires that shipping companies have a safety management system. That's like making a kid take a test that her teacher will never grade. What incentive does that kid have to do a good job?

The law says that every company has to have some kind of manual that details how crew and shoreside management should deal with various situations. The specifics of that manual—how it's written, what it covers, how it addresses each anticipated issue—are left up to the individual company. No one approves or evaluates it. The coast guard would like to be more proscriptive about the contents of these manuals, but there's a tacit agreement that any additional guidance would burden shipping companies.

In contrast, the International Maritime Organization (the branch of the UN dedicated to regulating shipping) issued an International Maritime Safety Code in 1987 in response to a spate of accidents caused by major management errors. This code, adopted by most of the world's merchant fleet (with the exception of the United States), offers clear SMS guidelines, has been continually tested and updated, and has resulted in safer shipping worldwide.

Before the hearings, members of the Marine Board asked TOTE how the company's staff learned about this SMS, and TOTE said that everyone aboard its ships was required to read and understand the manual. The coast guard pressed TOTE for more training information. After all, the SMS was useless if no one in the company was familiar with its standards and protocols. Especially in the event of an emergency, who had the time to digest a manual? So, the coast guard wanted to know, how did TOTE make sure everyone knew what was in that document?

Well, TOTE answered, we have a training video that the crew is required to watch. After much prodding, TOTE produced a VHS copy of said video. But they couldn't locate a VCR to watch it.

"I understand you're telling me that your safety management system is in compliance with the treaty," Odom said to Greene. "But did you have specific plans that identified specific emergencies? Did you have contingency plans based on known risks to your vessels along with plans required by the code to support the vessels from shoreside?"

"You know, I don't have them in front of me," Greene said. "I'm the president of the company. I have a staff that administers this."

Exasperated, Odom lost a bit of composure: "But did you guys ever have any discussions about the hurricane season? Not specific to the vessel, but in the company? Did you guys ever discuss hurricanes, upcoming hurricane season, and have any type of scenario, an exercise, training, or any of that with respect to the weather alert season?"

"We addressed the season last year by promulgating the safety alert that you've highlighted as an exhibit," Greene replied dryly. He was referring to a brief memo sent to all TOTE shipmasters at the beginning of the 2015 hurricane season reminding them that it was hurricane season.

Throughout his daylong testimony, Greene maintained that his employees were above reproach, that TOTE's mariners were the world's best, and that they complied with all national standards and regulations. But the evidence didn't support that. A month after *El Faro* went down, TOTE issued its monthly safety memo—a roundup of lessons learned in the past thirty days—to its staff. There was no mention of the cargo ship that had been lost under their watch. Apparently, there were no safety lessons to be gleaned from the loss of *El Faro*.

THE FOLLOWING DAY, CAPTAIN EARL LOFTFIELD, ONE OF THE MASTERS OF *El Yunque*, gave the public the first inside look at what it was like to work at TOTE. Like many captains, Loftfield was a passionate yet practical man. He loved the sea but understood his role in the larger shipping system.

Loftfield had worked for TOTE for more than a decade. He was the man who mastered *El Yunque* out of Jacksonville to Puerto Rico on October 2, the day after *El Faro* disappeared. The experience of steering through the sister ship's wake weighed on him heavily.

Loftfield was a literary man, drawn to poetry and proverbs. Shortly after his return from that voyage, he published a piece called "The Sister's Journey" on his union's website. It was there that he reprinted an email he sent to John Lawrence on October 4, 2015, at 10:30 a.m., as his ship encountered the last remains of *El Faro*:

"Latitude 23-23.910N, Longitude 073-57.451W. This is the apparent point of origin for plume of oil rising and creating a slick. At this location, oil was black on the water and air smelled strongly of same. We found the slick after traveling through a debris field for 25 miles, at times having as many as fifty simultaneous sightings of pieces of insulated containers."

On the following day, Loftfield gathered his crew in the mess hall of *El Yunque* to debrief. Of that event, he wrote, "Significance of what we have witnessed is acknowledged. The Pain. The Rage. The Knowing. Cautionary mention of predators ashore wanting to exploit the grieving and the possibility of 'hearing the truth you've spoken twisted by knaves to make a trap for fools.'"

Five days later, he held a memorial on the deck of *El Yunque*. "Crew gathered on the bow," he wrote. "Moonless night. Sea was flat. Eternity over the rail. With each of 33 strikes of the ship's bell, a flower was dropped in the water. Our ritual is complete. The mark on our souls will endure forever."

The heavens sent signs, which he recorded: "Lightning began far in the distance—two points to port and continued throughout the watch. A meteor burned bright, arcing towards the lightning. We sailors see what we see and have our judgments about what is indicated."

Loftfield addressed the Marine Board in measured tones and chose his words carefully. He said that he was deeply fond of the *Ponce*-class ships: "They made them long and skinny and aggressive looking, sort of like a barracuda. They're very attractive, with a very pointy front end and a very pointy stern. Like a '70s muscle car, which has a lot of steel on it." He was such a fan of the ships that in 2010, he invited John Glanfield and other former Sun Ship employees to tour *El Faro* when she was docked in Philadelphia. Glanfield was thrilled to be aboard one of his babies again.

But, Loftfield told the board, the ships had their limitations. Although TOTE's safety management system recommended that crew maintain stability with ballasting or by moving cargo, Loftfield said that that just wasn't possible at sea on the *Ponce*-class ships. Their design didn't offer a lot of ballasting flexibility, and you certainly couldn't shift around shipping containers without cranes.

Further, the ships' streamlined shape made them less stable in the water. To illustrate his point, Loftfield introduced the concept of "block coefficient," a ratio that defines a vessel's stability by comparing the shape of its hull to a solid box. The more you camber the sides of a boat's hull to reduce drag, the tippier it gets. Think of a flat-bottomed skiff versus a narrow and shapely kayak. A barge, which is shaped as much like a block as any ship (and, therefore, is very inefficient from a drag perspective), has 100 percent block coefficient, Loftfield noted. A racing sloop, with its sharply angled deep keel, has a very small block coefficient.

"Modern [cargo] ships have a much larger block coefficient than the PONCE class ships," Loftfield said. New ships, he said, look like a solid block of steel. The *Ponce*-class ships were all curves. "When the [*El Faro*] was tied to the dock," he said, "out of 790 feet, there's maybe 120 feet that actually touches the dock. For a barge, the entire length touches the dock." *El Faro* was designed for speed, not stability.

Loftfield also addressed the question of securing the scuttles. He said that chief mates would go down through the hatches to check the holds for water, but there weren't any scheduled rounds to ensure the hatches were secured. Once again, the responsibility to keep the ship watertight was everybody's job. Hence, nobody's.

Keith Fawcett, another Marine Board investigator, wanted to understand how TOTE operated after *El Faro* was lost. Did the company change at all? "Did they talk to you or communicate to you or discuss this at all with your sailing?"

"It was certainly on everybody's mind," Loftfield answered. But he made it evident that there was no formal response.

Then the captain expressed, better than anyone else could, some fundamental truths about the shipping industry: "In terms of sailing, it's what I do. It's what the ship does. It may seem cold and inhuman, but it's the job. It's why we're here. So yes, there's a casualty or there's a mystery, there's an unknown, it's not business as usual, but we've got to go on with business. The shelves in Costco and Walmart [in Puerto Rico] emptied out after the sinking of *El Faro* and we were the link—*El Yunque* was the link—supplying an entire population. There was a job to do."

In his mind, Loftfield could balance the horror of losing *El Faro* with the demands of his job. When he sailed to San Juan two days after the sister ship went down, he recalls: "We had the rather amazing experience of crossing right through the debris field with all hands on lookout and finding where the oil was bubbling up from *El Faro* and passing directly over it. And an eerie calm passed throughout the entire place. It was after we had finally passed the oil slick and debris field that we came back up to speed. We got to Puerto Rico as quickly as possible."

CHAPTER 27

PORTRAIT OF INCOMPETENCE

Pinpointing responsibility for *El Faro* became a shell game. At the time of the accident, TOTE Services managed the vessel, TOTE Maritime Puerto Rico owned it, and TOTE Shipholdings commissioned vessels; all were subsidiaries of TOTE, Inc., itself a subsidiary of a large, diverse logistics company called Saltchuk. All these separate yet related entities within the company obfuscated chain of command. That fact became obvious with the testimony of Tim Nolan, president of TOTE Maritime Puerto Rico. Nolan's division was still based in Princeton, New Jersey. He had joined the company in 2013. Like many of TOTE's executives, he was not a mariner; he had undergraduate and graduate degrees in marketing.

Nolan had met Captain Davidson a couple of times and liked him. When Davidson was being considered by TOTE Services to master one of the new LNG ships, he asked Nolan for a recommendation; in turn, Nolan encouraged Admiral Greene to grant him an interview for the job, which he did. Shortly thereafter, the mysterious administrative issue aboard *El Faro* came up, and Davidson's application was declined.

Although TOTE Maritime Puerto Rico was considered the actual owner of *El Faro* and *El Yunque*, Nolan and his staff didn't track their vessels. They left that responsibility to TOTE Services. During Hurricane Joaquin, Nolan didn't get any notices from that division alerting him that not one, but *two* of his ships were traversing the path of a major hurricane.

Tracking his assets wasn't Nolan's job. Instead, he was tasked with making money, making sure his operation remained profitable.

Tom Roth-Roffy was mystified by how all the various entities within TOTE operated. But more important, how did Nolan, ostensibly the ships' owner, make sure the management company was doing a good job? Was it possible that if they'd been wholly separate companies, he would have paid more attention to how closely that company was handling hiring, firing, safety, and ship repair?

The question caught Nolan off-guard. Why, Nolan asked in return, in thirty years would his company suddenly question TOTE Services's judgment?

Desperate for a road map that would clarify TOTE's chain of command, the Marine Board interviewed several TOTE executives and grilled them for information about who exactly reported to whom.

Jim Fisker-Andersen was trained as an engineer but had little sailing experience and only a third assistant engineer's steam license—the bottom rung in the ship's engine room. At the time of the accident, he was director of ship management for TOTE Services, responsible for overseeing all ship engineering and maintenance. Previously, he'd served as TOTE's port engineer in Jacksonville. He was promoted to his current position during the restructuring in January 2013. In his new role, he oversaw all of

TOTE's port engineers, including those who took care of the aging *El Faro*.

Fisker-Andersen was in San Francisco tending to problems with *Isla Bella*, TOTE's expensive new LNG ship, when *El Faro* sailed into Joaquin.

Port engineers were required to regularly evaluate ships' masters; the lowest-level engineer was in charge of judging the work of the highest-level deck officer. The two men worked in completely different orbits. One worked with machines, the other managed people.

Maybe that's why during Fisker-Andersen's tenure as director of ship management, Davidson's evaluations didn't get done. Fisker-Andersen may never have asked his port engineers to do the work. Time went by. Files remained empty. Regardless, the office people got promoted. If paperwork like evaluations failed to get done, Fisker-Andersen said defensively, it was the fault of human resources.

To whom did the ship captains report to? Fawcett asked him. *Uh, nobody*, Fisker-Andersen replied. The master was the master of the ship. He could use any of the resources ashore, sure, but he didn't *report* to anyone.

Who made sure the ships were being properly maintained? Who audited the port engineers? Not my jurisdiction, Fisker-Andersen informed the Marine Board. That would be the realm of TOTE's designated person ashore, Captain John Lawrence.

Some of the most bizarre, or rather, most confused testimony came from that particular gentleman the following day. His official title was TOTE Services's Manager of Safety and Operations, a job he'd taken in February 2014. He was the same man who had missed Davidson's first emergency phone call ashore and then was reconnected through the calling center for a single brief conversation, the last contact anyone ashore would have with the ship before it disappeared.

As the designated person ashore, Lawrence was responsible for fielding all questions from ship crews while they're at sea, and in that role he reported directly to Admiral Greene. For all intents and purposes, he was fairly high up the food chain at TOTE.

Lawrence, in his own words, oversees safety for TOTE. In short, his job is to implement the company's safety management system (SMS) and keep it up to date.

TOTE's SMS included a pair of manuals that provided procedures for the safe operation of *El Faro*, as well as a list of safety drills. The manuals included instructions about what to do in the event of loss of propulsion, flooding, and abandoning ship. Formally, weather wasn't identified as a risk to TOTE's ships. The one mention in the manual: "Severe weather is to be avoided where possible by altering the track of the vessel. Instruction for maneuvering in extreme weather can be found in 'The American Practical Navigator' HO Pub. #9."

Throughout Lawrence's mystifying testimony, the clearest message was that in the year or so that he had been working for TOTE, he'd passed most of his responsibilities off onto other people or departments.

"Yes, my responsibility is the safety management system for the entire company, for roughly twenty-seven vessels," he said to the Marine Board panel. "Let me go back—it's my responsibility to ensure the system is working as my position of a designated person. And to monitor the system and maintain the manuals. But again that's everybody's job."

How did he monitor the safety system and maintain the manuals? Audits, he said. Internal audits, in Lawrence's view, were about checking the master's paperwork. Lawrence delegated these audits to others in the company and hired third parties to perform external audits. He had auditors who audited the audits. Suffice it to say, there were plenty of people shuffling paper around, yet no one had ever audited any of the vessels' voyage plans or stability calculations as far as he could remember.

Lawrence's original job description said that he was responsible for evaluating officers, and certainly evaluating officers was required under the company's SMS, but he claimed that that task was no longer his responsibility. Whose was it? He didn't know. When a deck officer at TOTE was repeatedly caught sleeping on his watch, Lawrence declared that that wasn't a safety issue; it was an HR issue.

Although his position as the designated person ashore required certification by code of some sort, he said, he wasn't sure what that code was. He wasn't aware of "any actual certificate for DPA training," he said. "I don't have one myself. It's just basically the experience, but I'm not aware of a specific certificate." He didn't know whether certification was required of the person who filled in for him when he was away. (In fact, there was no such certificate.)

How many people in the office had DPA training? He didn't know. But he wanted to reassure the coast guard and the NTSB that TOTE works as a team. That statement was supposed to make everyone feel like TOTE had this whole vessel management thing under control.

During testimony, Lawrence wasn't able to describe the responsibilities of his fellow managers, and he couldn't tell the coast guard who in the company had the authority to defer maintenance on the ships. He also couldn't tell the coast guard exactly which of his duties had been delegated to which people within the company.

Although he was the designated person ashore for the captain and crew to contact in the event of emergency, he didn't notify the answering service when he was away, even when he knew he wouldn't be reachable. He was satisfied by the fact that his cellphone number was posted around the ships—folks could call him directly. Of course, that's assuming they had cell service. If someone had a problem while at sea, they'd need to ask the captain for permission to use the satellite phone. That would entail an awkward conversation if the purpose of the call was to alert shoreside personnel that the top officer was making a fatal mistake, like sailing directly into the path of a hurricane.

The call center number, which was to be used in a marine emergency or when crew members had other issues, such as human resources problems, was also posted around the ships. The people at the call center had been given the phone numbers of "eight or nine people" at TOTE. In an emergency, the call center operator would just start calling down the list until she or he got a responsible human being on the other end of the line.

It appeared that "safety" meant leaving a paper trail that all necessary drills had been done aboard the ships.

Monitoring tropical weather, however, was not part of the company's safety protocol. When *El Yunque* had issues with the mechanism that deploys the lifeboats, that wasn't considered a safety issue. When the Polish riders were doing all kinds of dangerous work aboard *El Faro* in rough weather, that wasn't considered a safety issue. Loading and stability of the vessel—that wasn't a safety management issue; that was a "safety concern," and it was the captain's responsibility. Monitoring or fixing the buttons and D-rings on *El Faro*'s deck to which cargo was secured wasn't Lawrence's purview either. That had been delegated to port engineer Tim Neeson, who only learned that he was responsible for that work during the Marine Board hearings.

"Were you aware that the vessels were operating close to their [Plimsoll] marks on the TOTE run in Puerto Rico?" This question for Lawrence came from an NTSB accident investigator on the panel questioning witnesses, Mike Kucharski. He was a salty former ship captain from New York with several decades' worth of sailing experience—someone who'd actually mastered *El Faro* when she was the *Northern Lights*.

"No, sir."

If a master had a concern or question about stability or routing, who should he go to? Lawrence gave a vague answer—anyone in the office could help steer the master to the right person, he said. Maybe the port engineer would know. "Again," he said, "it's kind of a team sport. If we don't have the answer, we find the answer."

If a master had a question about his routing, say when a storm was forming in the Atlantic, who should he turn to?

"You got a lot of expertise in our office," Lawrence said. "Combining hundreds of years of experience whether that be my maritime background for 40-plus years . . . I'm positive we could address any questions the vessels bring to us."

TOTE also lacked vessel-specific heavy weather plans—and provided no risk analysis or training for any of its staff, officers, or crew at the commencement of hurricane season.

"It's the master's responsibility for preparing for a storm and for doing his voyage planning . . . we don't tell the masters what to do," Lawrence reiterated.

In spite of Lawrence's supreme delegating skills, an external audit conducted the year before the accident recommended that TOTE hire an additional person to assist him in his duties. Instead, the company opted to shift more of Lawrence's responsibilities onto other people already working there.

A picture was emerging of a disconnected and distracted executive team and an overburdened staff that actually did most of the work.

The weekend before the accident, Lawrence was on the West Coast for a wedding. Then he traveled to Atlanta for an industry meeting and returned to his home in Jacksonville on Wednesday evening, September 30. The next morning, he missed Davidson's emergency call. While he was traveling, Lawrence had not delegated anyone in the company to fill in for him.

It was during Lawrence's travels that Davidson sent the long, carefully reasoned email to his safety officer requesting permission to return through the Old Bahama Channel because he had concerns about Hurricane Joaquin. Lawrence did not reply to Davidson's request.

Why not? "Because I was surprised [Davidson] was asking permission," he said. "I saw no reason to answer at that time because we still had a couple more days before he would even be starting that route."

The email, at the very least, should have alerted Lawrence to the fact that a hurricane was out there in the Atlantic, somewhere in the vicinity of his ships. But after receiving that email, the safety manager of TOTE Services didn't bother to check the weather. That's why his mind was a total blank when he got Davidson's urgent voice mail message on the morning of October 1.

(Lawrence also didn't know that the emergency calling center's code "PC" meant "Please call." *Did the call center have a log of their acronyms that you might need to know?* "They may but I've never asked for it," Lawrence replied.)

When he finally talked to Davidson that morning, Lawrence said he knew the hurricane was somewhere in the Atlantic, but he wasn't aware that *El Faro* had sailed anywhere near it. Neubauer pointed out that if Lawrence had read Davidson's email sent the day before, then he would have known of the captain's plan to go south of the hurricane by a mere sixty-five miles. That in itself should have set off alarm bells.

"Yes, sir. Yes, sir," Lawrence answered helpfully. "I was interpreting your use of the word 'vicinity' as meaning *right on it.*"

Lawrence said that when he got the emergency call, he didn't infer that the ship's proximity to Joaquin might have caused an issue. "I knew [Davidson] was having some type of problem that he wanted to contact and talk to me about. At that time, I did not, you know, put the two together that that was the reason for his call."

Davidson had told Lawrence that he had "free communication"

with water and a "pretty good list." Why didn't these terms worry Law-
rence? Why didn't he go into full-on search-and-rescue mode?

"I really didn't speculate on the 'pretty good list,'" Lawrence
said. "Again, [Davidson] seemed very calm through the voice message, so I didn't speculate until I talked to him and found out more
information."

In order to get Lawrence on the line, Davidson had had to
make a second call to the emergency call center, and that conversation was recorded. The audio of that call was played at the hearing and sent a chill through the room. Some people left the room
before it was broadcast because they didn't want to hear the voice
of a dying man. You could clearly hear the panic in his voice. "Oh
man," he said to the operator. "Oh God. The clock is ticking."

Lawrence had a ten-minute conversation with Davidson after that call, but the conversation wasn't recorded—the emergency
call center had patched him through.

Why did Davidson express so much urgency to an anonymous op-
erator and then compose himself when he spoke to TOTE's man ashore
a minute later?

The NTSB had interviewed Lawrence before—he was one
of the first witnesses the agency spoke to in Jacksonville shortly
after *El Faro* disappeared. Even then, when the scene was fresh,
Lawrence didn't quite have the ability to piece it all together. He
showed up to speak to a roomful of investigators from the coast
guard and the NTSB, people who desperately wanted to understand the tenor of that final phone call with the captain of a sinking ship.

It was Lawrence's moment in the spotlight. He'd been a safety
manager for fleets of ships for much of his career, safety was number one, but now he couldn't remember basic details about his brief
conversation with the doomed captain of the one ship he'd just
lost. Lawrence also failed to bring the notes he'd made that tragic
morning.

Instead, he gave vague answers to the investigators' answers and continually reminded them that he was lost without his notes.

He didn't know whether *El Faro* was on emergency power from the diesel generator. He wasn't sure whether the ship had lost power or propulsion or both.

It didn't take long for the investigators to show their frustration both about his foggy memory and the vague questions he'd apparently asked the doomed captain during that single critical phone call.

"So based on your conversation with [Davidson], were you able to form an opinion as to how dire the situation was at that time?"

"I think the only thing I can really say at this time is that his demeanor, it was very similar to the voice message he left. He seemed to be at the same calm level as that throughout the phone call. Very businesslike, matter of fact. He did not seem to panic."

The group reconvened later that day and this time Lawrence brought his notes. But those chicken scratches, scanned by the NTSB and folded into the permanent record of the investigation, contained woefully little information. Lawrence had recorded *El Faro*'s position, the fact that a scuttle had opened, and that they had considerable water in the number three-hold and no main engines. He recorded that they had a 15-degree list and ten- to fifteen-foot swells. They were trying to pump out the hold, but, Davidson said, they didn't plan to leave the ship. That said, they were setting off their alarms to alert the coast guard.

If a captain is at sea, in the middle of a hurricane, he's lost propulsion and is taking on water, and talking about whether or not to abandon ship, doesn't this sound like an emergency to you?

Lawrence didn't know. He couldn't say. TOTE's man on the ground responsible for keeping the crew of twenty-plus ships safe had no idea *El Faro* was about to go down.

"I did not feel an immense emergency," he said. "What alerted me a little bit was the one sentence he said, 'We do not plan on leaving the ship.' That did make an impression on me that maybe this is a little more serious than I was thinking at the time. . . . He was calling me to let me know that he was going to be pushing emergency buttons, to give me his position and letting me know that he was not in a grave situation and that he was going to push his distress buttons."

A few minutes after he hung up with Davidson, Lawrence got a call from Chancery at the coast guard, and together they agreed, based on Davidson's tone, that the ship was in a disabled phase, not a distress phase. "So what [the coast guard] said, reading between the lines [of my notes], they told me that they wouldn't be sending anybody out at this time because it's just disabled."

He didn't push them on the idea that the ship might be in dire straits; instead, he was relieved that the coast guard was assessing the situation and would take it from there.

On the morning of October 1, a few people at TOTE's Jacksonville office plotted the hurricane and *El Faro*'s last known position. *El Faro* was at the eye. "That's when they started putting the pieces together," Lawrence said.

When the VDR was finally recovered from the seafloor, the public found out exactly what Davidson told Lawrence during that conversation. His side of the story was caught on tape and carefully transcribed, then posted online.

From the dead man's words, anyone could see that *El Faro* was in great peril.

CHAPTER 28

MISSION NUMBER TWO

On the USNS *Apache* in late October 2015, the CURV-21 proved a clumsy tool for searching in the sunless deep three miles down for something the size and shape of a coffee can. In short time, the CURV's navigation system stopped working entirely. Its narrow cone of visibility, further limited by the short range of the vehicle's spotlight, rendered the search time-consuming, and ultimately, in vain.

Without navigation, the techs had to rely on dead reckoning to maintain their bearings. Using the main wreckage as a starting point, they drove the CURV in a radial pattern, always returning to the half-buried hull to reconfirm their orientation. As the vehicle crept through the city of junk, it used two sonars—high and low frequency for high and low resolution—to scan for anything metallic. There was a lot of metal down there, and most of it was shipping containers. Everything that the CURV detected had to be investigated.

Obstacles would suddenly emerge from the darkness, blocking their way, demanding quick course changes to avoid a collision.

The team often got disoriented in the murky dark. They'd drive the CURV far out from the main wreckage, or so they thought, and then come back upon it sooner than expected. It was clear to Tom Roth-Roffy that they didn't know where they were. His worst fear was that they could breeze right by the VDR without knowing it. The whole endeavor seemed futile. There had to be a better way.

After days at sea, the waves began to kick up, rocking the *Apache* enough to make people feel ill. Morale aboard dropped. The tech team wanted to go home. Debilitated by seasickness and dopey from Dramamine, Tom refused to quit and called the NTSB headquarters to tell his superiors as much. Headquarters agreed to enlist the help of their own research and engineering specialist—Dennis Kryer—in the search. He had a gift for finding things.

Kryer asked the *Apache*'s tech team a lot of questions. He wanted the mass of the wreckage so that he could build a computer model of how the ship went down. He carefully studied the splash *El Faro* had created on the seafloor to determine its speed and direction when it hit, and he calculated the buoyancy characteristics of the hull and the house to predict their drift pattern. He used this information to come up with coordinates for the VDR's location. "As it turns out," Tom says, "he was pretty close."

Kryer recommended that they spend another few days extending the search much farther to the north of the hull. After a quick trip to Puerto Rico to refuel, the *Apache* returned to *El Faro*'s position, took one last look, and finally located the bridge 450 meters off the port bow.

If the VDR was still attached, that's where it would be. But the NTSB's $1 million contract with the navy was up. The *Apache* team carefully hoisted the CURV on its three-mile-long fiber-optic umbilical cord back onto the deck and headed back to Virginia.

Had Captain Davidson been pressured by TOTE? Was he missing critical weather information? Was he suicidal? Without the VDR, investigators could only speculate.

But that wasn't Tom Roth-Roffy's style. If there was a way to get the VDR, even if it took extreme measures, the NTSB would do it.

With the support of Congress, the NTSB received an additional $1.5 million to pursue a second mission in the Caribbean in April 2016.

Tom Roth-Roffy stayed behind, leaving the job to NTSB's Eric Stolzenberg.

Stolzenberg began his career as an engineer in the merchant marine. He shipped around the world on supertankers and eventually got a land-based job as a naval architect—he'd studied the field as an undergrad at New York Maritime College—before joining the NTSB in 2008. Now in his early forties, he was generally younger and brasher than his NTSB counterparts. Tall with a head of thick brown hair, Eric played Will Smith to Tom's straight-shooting Tommy Lee Jones.

Stolzenberg watched the *El Faro* investigation from the sidelines for a couple of weeks, eager to jump in. He loved solving mysteries that required piecing together obscure clues. He was also drawn to the incident because, he told me, hurricanes fascinated him. He first got hooked in 1985 when Hurricane Gloria made a direct hit to Long Island, where he'd spent summers with his family.

When the wreck of *El Faro* was located in November, Stolzenberg was designated the lead on the NTSB's naval architecture team. He immediately began acquiring any extant *El Faro* blueprints, as well as her loading diagrams and photos. He was sure that *El Faro*'s design played a major role in her undoing.

Stolzenberg's quest was to determine the sea conditions in which *El Faro* sank and use that information to create a dynamic model of the events leading up to her demise. He stepped onto *El Yunque* in December and homed in on her downflooding points. "A ship has an enclosed buoyancy," Stolzenberg tells me. "When all the hatches are watertight, a ship remains buoyant. You cannot sink it. Physics dictate it can't go to the bottom. You've got a balloon in there keeping it afloat. Poke enough holes in that balloon and the inevitable occurs."

On *El Yunque*, he could see how those vents, at a certain angle of heel, would get submerged.

One big question: Did this design meet current codes, and if not, why?

When *El Faro* was converted from roll on/roll off to a container ship in 2006, the regulators—the US Coast Guard and ABS—acquiesced to TOTE's repeated requests that the work not be deemed a major conversion. Accordingly, the regulators did not redo *El Faro*'s damage stability index. "And that might be an oversight," Stolzenberg says.

A damage stability analysis determines how many holes you need to poke into a ship to make it sink. The calculation is based on probability—where a ship is most likely to get hit—and dynamic modeling.

Stolzenberg says that determining a ship's damage stability is both science and art. But at least it gives you a sense of a given vessel's vulnerabilities. This would be critical information for a ship's captain when he's thinking about heading into a large storm.

Just like John Glanfield, Stolzenberg saw the vents for what they were: ways for water to compromise *El Faro*'s buoyancy. The ship's designers saw that, too, and equipped her with fire dampers—big steel louvers that could be closed to snuff out a fire. The dampers weren't weathertight, but if closed, could slow the ingress of seawater during a storm. Crews on *El Faro* regularly tested the operation of these dampers during fire drills but never considered using them to protect the vessel in high seas.

If a loose car floating in three-hold had hit the fire pump causing a breach in the hull, the ship would have flooded very quickly.

Stolzenberg thought a lot about these various scenarios and wanted to get a closer look at the wreck. He also wanted that VDR.

In April 2016, he boarded the R/V *Atlantis*, an oceanographic research vessel operated by the Woods Hole Oceanographic Institution, owned by the US Navy. Two undersea vehicles they carried with them would work in tandem to document the accident site and, with luck, locate the VDR. The automated underwater vehicle (AUV) was a tethered machine with broadband sonar that ran along a preprogrammed path, identifying anything that approximately fit the size and shape of the ship's mast. Its twenty thousand feet of fiber-optic cable alone probably cost $1 million.

The SENTRY, another autonomous vehicle, was equipped with searchlights and high-resolution cameras. It followed behind the AUV and was programmed to investigate each target the AUV identified. Unfortunately, the AUV's margin of error was thirty meters, so often it took a while, sometimes as much as forty-five minutes, for the SENTRY to find one target.

The AUV and the SENTRY could never be in the same area or they risked crashing into each other, so sometimes the tech team sent the SENTRY on a wild goose chase far away from the scene while the sonar worked closer to probable sites.

As before, the team aboard divvied up the watches, monitoring the computer screens in the nerve center to see if they could detect the mast. But they were stumbling in the dark.

The techs from Woods Hole Oceanographic Institution who ran the robotic equipment sat down with Eric as they were heading out to the accident site. They said, *Remember: At fifteen thou-*

sand feet, there's no guarantee you're coming here tomorrow or the next day. Things go wrong down there. Every photograph, every dive, every view could be your last, so treat it as such.

During the first couple of days, Stolzenberg got a good taste of what they were talking about. Fittings on the vehicles' casings leaked, small things went wrong, and Stolzenberg quickly learned that, *Yeah, this could be the last time we see things.* The autonomous vehicle took two hours to sink down fifteen thousand feet, and two hours to surface, which meant four hours of waiting each time they brought it up to upload and process its data. Time gnawed away at everyone's patience. They only had six days over the wreck to find the VDR, and everything was an exercise in frustration.

The sonar picked up countless targets—automobiles, shipping boxes, and car batteries—resting quietly on the seafloor. It identified personal effects, standard household goods, bicycles, toiletry items, dolls, and roof flashing, all eerily still in the deep. The mast was big, about as big as a container. All the ship's containers had fallen off as it went down and were now showing up on the sonar as nice, bright targets. There was an impossible amount of stuff to go through.

While the SENTRY investigated each of these targets, small, strange-looking deep-sea creatures wandered in front of its high-resolution lens. Three miles down, it was another world, now littered with the detritus of terra firma.

Wreckage distribution told Eric a lot about how *El Faro* fell through the water. From the wheelhouse north was mostly cargo; from the wheelhouse south to the hull was ship debris. The first mission identified where the house was, so, he concluded, the mast must be somewhere between it and the hull.

With four days left, hundreds of possible trails to explore, and two hundred identified targets to investigate, Stolzenberg threw a Hail Mary. They had to limit their search to the area between the hull and the house.

As soon as they committed to Stolzenberg's plan, the trail got hot. They found railing from the wheelhouse, the ship's stack, and a piece of steel with a porthole, all of which told them that they were in the right place.

They picked through the trail of debris, investigating everything they encountered. That night, when his shift was up, Stolzenberg refused to leave his post in the tiny control room. He'd been up for twenty-four hours but couldn't pull himself away from the monitors. He could see maybe three or four meters in each direction, and a lot of "snow"—tiny particles that got caught in the spotlight—but he knew they were getting close.

A few minutes into the next watch, the SENTRY missed yet another target and had to swivel and go back. This time when it turned around, one of Stolzenberg's team members saw a flash of light to its right. What the hell was that?

They steadied the robot and slowly approached the area from which they'd seen the flash. "Then," according to Stolzenberg, "we were like, holy moly. That's the reflective tape on the VDR!"

The superreflective tape wrapped around the VDR canister was no wider than two inches, but it bounced the high-intensity lights of the underwater search vehicle back at its camera. Just beyond the VDR, Stolzenberg could make out the three legs of the mast coming at right angles to the SENTRY; the rest was buried in mud.

"I guarantee you that on the first *Apache* mission, Tom went right by this thing," Stolzenberg says. "You have to be so close to see anything down there with that much junk around. You almost have to be on top of the stuff."

The tech team didn't want to get the SENTRY's tether tangled up in the mast, so they approached with caution and documented the scene. At 2:00 a.m., when they were certain that they'd found the VDR, they sent an email back to the NTSB.

The next day, they deployed the autonomous vehicle with its

high-resolution lens to document the entire site. During that mission, the SENTRY took more than forty thousand photos, which the NTSB later patched together to create a map of the scene.

Meanwhile, they had to figure out how to extricate the VDR from the mast's clutches. The steel canister was held to the mast with two quick-release bands. Although it sat proud, you couldn't just scoop it up. You'd need a highly skilled robot to spring it free, drop it in a basket, and carry it three miles to the surface.

Stolzenberg didn't have that kind of robot with him on the *Atlantis*. He could only stare at his prize, sitting undisturbed on the ocean floor in the watery deep.

CHAPTER 29

THE PROOF IS IN THE PUDDING

With or without the VDR, the official inquiry continued. The Marine Board of Investigations met again in Jacksonville for a second round of hearings in May 2016. Many of the witnesses brought forth were experts in their fields, brought in to discuss the more technical aspects of weather reporting, stability calculations, inspections, and the cargo-loading software.

But two witnesses stood out: Captain Jack Hearn and Peter Keller, the executive vice president of TOTE. The testimony of these two men provided the starkest contrast between those who toil in the trenches and those who call the shots. One had made his living on the high seas for four decades, working with nature, machines, and the rough folks who sometimes end up on ships. The other ran various corporations from offices, boardrooms, conference rooms, and golf courses. It was as if they were from different countries, speaking different languages.

Yet both were critical to the modern maritime industry.

Captain Hearn had served as a shipmaster for twenty-five years, seventeen of those years as master in Alaska on the *Northern Lights*, the ship that would become *El Faro*. In 2007, Hearn was transferred from Alaska to helm the *El Morro* on the Puerto Rico run.

After complaining for years to TOTE about the *El Morro*, that vessel was eventually scrapped, and *El Faro* took over her run. But for Hearn, it was a Pyrrhic victory. Shortly after he was reunited with his beloved ship, the seasoned captain was forced to resign when it was discovered that some of the unlicensed crew had transported drugs on his watch.

On Hearn's last day with *El Faro*, a new captain came aboard to take the helm. His name was Michael Davidson. Hearn was given less than an hour to download nearly two decades' worth of knowledge about the vessel to Davidson. Then he was hustled off the ship.

Four years later, Hearn was still bitter about his treatment at the hands of TOTE. His "firing" gnawed at the fiftysomething father of five. He'd served his company well, more than well. Over such a long period of time in such a challenging environment, away from his wife and family, he'd maintained a practically flawless record of professionalism and good judgment. The idiots ashore had sullied an outstanding maritime career.

At the Marine Board hearing, Hearn didn't mince words. In 2012 or 2013, he said, a number of shoreside positions had been eliminated, leaving incompetent staff onshore to run operations.

From Hearn's perspective, operations manager Don Matthews was a poor substitute for Bill Weisenborn, the former manager with whom he'd worked until TOTE's restructuring. Weisenborn had graduated from the prestigious US Merchant Marine Academy at King's Point; he was an experienced mariner who'd soaked up invaluable knowledge working with multiple shipmasters over the years. That made him an excellent shoreside partner, and Hearn had worked closely with him on voyage planning and weather avoidance throughout his tenure at TOTE. Weisenborn even sent extra weather reports to the ships when they were out to sea. He cared about his people, even when they were out of sight of shore.

Before the restructuring, Hearn said, "If you changed your route, there would be a discussion and a joint decision because you just don't change from the routine without advising and sometimes getting advice from the company about the need for the mission of the voyage. That's how decisions were made. With the new group, I wasn't always sure exactly who was making the final decisions."

Matthews, Hearn announced to the room, did not have the marine background to be useful to an experienced seaman like him. Matthews lacked voyage and operational experience with the ship; Hearn would never seek help or advice from this man.

Family members sitting behind him cast knowing glances at each other. This was exactly the kind of testimony they'd been waiting for. They thought TOTE could be held accountable for downsizing to the point where necessary expertise was lost.

Hearn also shed light on why Davidson might have ignored the advice of his mates. Hearn said that as a captain, it takes a long time to develop a solid working relationship with a chief mate. You're the one person in charge of a huge vessel, the cargo, and all the people aboard. It's a tremendous responsibility. You're not just going to hand over your trust to the person standing next to you, even if he or she had fancy credentials.

"It's a year or two before a guy really gets the routine down of voyage experience, weather experience, seasonal experience, and expectations," Hearn said. "It takes time. It's not something that happens immediately."

A captain might be even less inclined to listen to his chief mate, and more inclined to rely on his own instincts, if the captain thought his job was on the line. The stakes would be too high.

Hearn said that after sailing on *El Faro* for nearly two decades, he was very aware of her vulnerabilities. She had lots of openings that could cause flooding in bad weather, plus, she had a low free-board, so he knew to avoid bad weather whenever possible. Her old steam plant demanded experienced people, the kind who knew the complex system so well that they could quickly deal with a situation if something went wrong. And things were always going wrong.

Toward the end of his time at TOTE, Hearn became increasingly concerned about the *Ponce*-class ships' ability to handle the demands made on them. "We were going to full loads," he said. That changed how the vessels handled. "The ship was very tender," he said. *El Faro* righted herself more slowly. "You could even feel the ship list—lean over—as she rolled from a rudder command alone, let alone rolling with a heavy swell. There was always a concern that she wasn't going to right herself adequately in other conditions."

This apparent lack of stability worried Hearn and TOTE's other experienced masters enough that in 2011, they convened through Bill Weisenborn to establish a margin of safety for the fleet of old ships, based on load calculations, below which they agreed they wouldn't sail out of Jacksonville.

Hearn's testimony ended on a sober note. As a captain, Hearn told the board, he always sought out a gentle ride. "You're always concerned about every detail of the ship and you want to preserve the original stability and the original position of the ship as long as possible. Not only for cargo equipment, but personnel, efficiency of the voyage, maintenance, the work that's being done, people who are working around other components. Things that are hot, tools, steam lines and the galley, guys are moving equipment they could get hurt. So you want to keep the ship as stable as possible for the personnel living on board."

There was a lot of shaking heads among those who'd lost their loved ones to stupidity or avarice. Now, the whole accident seemed completely preventable. If only wiser and more experienced people had been in charge.

IF THE TERM "CORPORATE GREED" WAS ON EVERYONE'S MINDS, THEY WERE about to get a fantastic display of corporate hubris. Executive vice president of TOTE, Peter Keller, the mastermind behind TOTE's restructuring, lumbered over to the witness table. He was older, maybe seventy, and very tall, which gave him a hunched appearance when he leaned into the microphone to answer the Marine Board's questions.

At the hearing he made it very clear to all that in his long, dazzling career, he'd mostly focused on the business end of shipping—restructuring companies, managing bankruptcies, consulting, overseeing acquisitions, and managing labor relations. He came across as an important man who didn't have time for the niggling details of ship operations.

Keller had been consulting for various entities when, in the summer of 2011, he was approached by the American Shipping Group (a division of Saltchuk Resources) to look at one of its companies, then Sea Star, now TOTE Maritime Puerto Rico. "The company was not functioning effectively," he said. "There were major issues with senior management in my view. The company was not properly organized. It had far too many people. It was overburdened."

In a couple of months, Keller had, in his own words, developed a plan to "reorganize and redevelop that company." It wasn't actually a plan so much as a feeling—just bullet points on a piece of paper. Shortly thereafter, he was handed the reins of Sea Star to execute his vision.

Keller set about removing what he deemed redundancies within the company. He "changed out" most of the senior management.

He "consolidated" the knowledge base in maritime operations. He says there were ten people in Jacksonville to oversee three vessels, and just as many in Alaska and New Jersey. "We knew we had too many people," he told the board. "We had too many people," he repeated.

"At no time," he told the board, "did we ever not have competent people in both Tacoma and Jacksonville to work with the ships. . . . And at the end of the day what we did is we put all that together into a more cohesive, better managed organization."

The following year, Keller ordered the construction of the new LNG ships, assigned the leaner staff with the task of overseeing the launch of these new ships, and moved up TOTE's executive ladder. To manage crewing and ship operations, he created TOTE Services. It was a "consolidation of resources," drawn from staff once based in Cherry Hill, New Jersey; Tacoma, Washington; and Jacksonville. All three of these offices were combined into one central group, he said, that was eventually moved to Jacksonville. Bill Weisenborn and many others declined to relocate. The crewing manager, for one, didn't mind making the move.

TOTE Services, as far as Peter Keller was concerned, was "very effective," "reliable," and "good."

He lectured the Marine Board on the wisdom of downsizing: "One of the things that happens in organizations if they become too bloated—and I think we all know this—after you do have some reduction in staff, the organization will work better. That was exactly the case with Sea Star Lines," he proclaimed. "And the proof with anything we do is, as they say, in the pudding."

Keller's arrogance pushed Tom Roth-Roffy's patience to the brink.

During the proceedings, Tom had thought long and hard about his NTSB career. He was a dedicated civil servant, a careful man, a thoughtful investigator. He saw people less skilled than him promoted out from under him, Peter Principled. The meritocracy didn't hold. A few months before the sinking of *El Faro*, he'd applied to SUNY Maritime to run their training ship, a fifty-five-year-old steamship called the *Empire State*.

Now, right before his eyes, in the testimony of Peter Keller and the rest of TOTE's executives, he saw the worst of corporate America. He'd heard enough. He carefully prepared a line of questioning for Keller, then read it verbatim.

"Tom Roth-Roffy, NTSB. Good afternoon, Mr. Keller," he began.

"Good afternoon, sir."

"Just recalling what you said earlier about the work that you initially started when you started working with Sea Star. I believe you were in kind of a consultant role to evaluate the performance of the company and to make recommendations on restructuring."

"Yes, sir."

"And, sir, do you recall if there were any formal reports that you had written or others had written that had documented the analysis and the recommendations that led to the eventual restructuring of the companies?"

"Most of it was verbal. There was one note that I remember sending that outlined eight or ten or eleven points that needed to be done from my view. Including things like the Philadelphia service and changing systems and process and people, things of that nature. But it was more verbal."

"So there was no actual consultant formal report that would lay out the issues, the rationale and analysis behind the recommendations?"

"No, sir."

People lost their jobs. Good people with a high degree of competence had been thoughtlessly purged. There'd been no formal plan. It was slash and burn.

"And then recalling further what you stated earlier this morning—and I'm paraphrasing—the evidence of a well-run operation are the results. Further, you stated that to have good results you must operate safely among other things. And as you said, *the proof is in the pudding.* To me that means that the end result is the mark and the success of the company's leadership and management. Would you agree?"

"Yes," Keller answered.

"In addition," Tom continued, "you stated that if you have a breakdown in an operation, there's probably some reason for it. And that the leadership of the company takes responsibility. Would you agree with that summary?"

"Yes. I said that, yes."

"Now, sir, many would argue, and few would dispute, the loss of the ship, *El Faro,* and its cargo and most importantly, the loss of thirty-three souls aboard *El Faro* represents a colossal failure in the management of the company's responsibility for the safe operation of *El Faro.*"

At that point, two of TOTE's lawyers leaned back in their chairs and glanced at each other.

"As you stated, *the proof is in the pudding.* And, sir, you have no doubt thought long and hard about the nature of the management failures that led to the loss of *El Faro*'s crew. Could you please share with this board your thoughts about the nature of the management failures that led to the loss of *El Faro*?"

The room went cold. Family members sat in their chairs holding their breath. Could it be that, finally, someone on the board was saying what they were thinking? All this time, it seemed that much of the proceedings was a formality and that, ultimately, all blame would lie at the dead captain's feet.

Now the lead investigator of the NTSB was demanding that the corporation, and its hatchet man, take responsibility.

Their eyes all turned to the back of Keller's slicked-back gray hair, which was all they could see, as his head emerged like a turtle's from the dark blue suit. He leaned into the mic.

"I think this tragic loss is all about an accident and I look to this board, as well as the NTSB, to try to define what those elements may or may not have been," Keller replied slowly. "I, for one, with fifty-one years of experience in transportation, cannot come up with a rational answer. I do not see anything that has come out of this hearing or anything else that I've ever seen that would talk about a cause. Certainly as management we look for that. We look for what NTSB and this board may come up with. Because we think it will be important. At this point in time, I, for one, cannot identify any failure that would have led to that tragic event."

That was it. His tone was either supremely arrogant or profoundly cruel. Interpretation depends on how you feel about those who lost their lives on *El Faro*. And those desperate for answers.

Many of the family members returned to the hearings that afternoon holding cups of pudding. Now they had a rallying cry. And a hero.

The next morning when Tom Roth-Roffy arrived early at the Prime Osborne Convention Center for the final day of the second round of hearings, he walked into the large room, sat down at his place on the panel, and started looking through his notes.

Captain Neubauer walked up to the proscenium and beckoned him over. "Come with me," he said. "These guys have to see you."

Tom followed Neubauer to the small room, almost a closet, that separated the hearing room from the grand hall beyond. Inside, four male lawyers, all representing TOTE, stood waiting for him in their dark suits. They glowered at the petite federal agent, then circled him wolfishly. One by one, they challenged his line of questioning the day before. *It was out of line*, they said. *Up until yesterday your questions were reasonable, but senior leadership at TOTE had lost a lot of sleep over what happened yesterday afternoon. We are thinking of writing a letter to Congress to protest your line of questioning.*

In all his years as an investigator, Tom had never been confronted by corporate lawyers as aggressive as this. There were rules. There was civility. There was mutual respect. Tom represented the best of government—an independent organization charged to uncover the truth. He was nonpartisan, honest, selfless. Lawyers always respected his authority. They could object to a question during hearings or behind closed doors, but attacking him this way was unprecedented.

"Whoa," he said. "Guys, I did not try to . . . my line of questioning was not intended to be accusatory or draw a conclusion."

But maybe it was. He started backtracking. He offered to make an apology. "Because with four attorneys threatening me to write a letter to Congress, it made me step back and say, *Am I in trouble now? Did I overstep, out of bounds, asking the question that I did?*" His biggest concern was his professional reputation. Even if he quit, a federal inquiry would be a blot on his spotless record.

I met Tom at SUNY Maritime in his stateroom on the *Empire State*, eight months after he resigned from his post at the NTSB. He told me that he knew how Congress worked, and at the time, he couldn't imagine who the lawyers would have written a letter to. But the encounter scared him. In his eighteen years at the NTSB, he had always been treated respectfully, as the truth seeker he was. The world, however, was changing. Facts were no longer just facts. Everything was debatable. Playground bullies in expensive suits had stormed the gates.

The lawyers had handed him a draft apology, which he reviewed and modified a bit. "That's what I read into the record on the last day," he told me. "And that was the last I heard of it."

Tom said that he had not intended to defame TOTE by the question, and, in fact, he hadn't. But the lawyers had discovered his weakness: his obsessive professionalism and sense of propriety, and they used it to scare him into submitting to their demand for a retraction. "They said management—senior management—last night lost sleep over this. And I defamed them. I don't know if that's the word they used, but I tarnished them by the way I'd asked the question."

He was still mulling it over in his head.

"They lost a fucking ship," I said.

"Yeah, that was my point," Tom said, suddenly empowered to speak honestly. In a matter of seconds, after years of being careful, he cast his stoic federal investigator mantle to the winds.

"You know the proof of the pudding?" Tom said to me, as if he was back in that tiny room, confronting the lawyers. "You guys lost a ship! Something was done wrong. I'd given TOTE an opportunity to say, 'Yeah, we're modifying our management structure. We're adding this position. We're conducting our own investigation. So far we've done this.' But you know, their response was nothing like that. After all the hearings and the tragedy, they still stood there and said, 'Nope. Nothing is wrong.'"

CHAPTER 30

VOICES

Somewhere in a closet in Washington, DC, sits a shoebox-size red plastic bin. A typed paper label reads: "Accident #DCA16MM001. Operator: TOTE Services. Accident date: 10/01/2015." Inside is what looks like a piece of junk, a nondescript steel canister the size of a toaster. Someone affixed a red tag to one of the canister's handles with wire, like something you'd find hanging off the toe of a cadaver. A cable hangs out from the top of the canister, its end severed and frayed.

This cable once led to the microphones on the bridge of *El Faro*.

The thing may look completely unremarkable, but the story of how it finally landed here is.

In August 2016, a team boarded the *Apache* once again to recover the VDR, this time at a cost of $500,000. Once again, they sailed out beyond San Salvador and idled the navy vessel above *El Faro*'s final resting place. Once again, they used the *Apache*'s powerful winches to lift the bright yellow CURVE-21 off the deck, swing it over the side, and gently lower it into the ocean.

Once again, the vehicle's navigation system quickly failed, leaving them to rely on dead reckoning as it picked its way through the mountains of debris on the ocean floor.

Fortunately, the NTSB had created a map of the accident scene by stitching together forty thousand photos of the wreckage taken during the two previous undersea operations. The CURV's operators—contractors from Phoenix International, a marine services company—used this digital tableau to navigate the vehicle, equipped with a small metal basket, safely to the VDR.

When it touched down, the CURV puttered lazily toward its target, continually checking its surroundings to make sure that its fiber-optic umbilical cord wasn't getting tangled in the junk all around it. This maneuvering took a couple of hours—there was no room for unforced errors.

The CURV operators aboard the *Apache* were remarkably skilled at manipulating the robot's Swiss Army–like appendages; they could use their shipboard laptop console to get the thing to tie a knot into a string. They drove the CURV toward *El Faro*'s twisted mast—now a helpful guidepost in the murk—where the VDR had patiently waited for them for eleven months. When the CURV approached its prize, it set its basket down beside the VDR, kicking up fine particles that dusted everything in the deep, temporarily obscuring the operator's vision. It waited for things to settle back down. Then it got to work.

Two quick-release clips held the steel canister to the mast. The CURV's robotic pincers reached out and delicately flipped up the clips. Click. Click. Then it gripped the VDR's two steel handles and pulled the canister out of its cradle. The VDR was free.

The CURV lowered its precious bounty into the basket, closed and secured the top, and gingerly retreated back to the surface with its prize, like a shopper returning from an underwater market. Doug Mansell, an NTSB engineer, leaned over the deck of the *Apache* on the night of August 6, squinting into an ocean illuminated by the vessel's spotlights. In a few hours, the CURV emerged from the deep.

It was hoisted aboard and relieved of its basket. Finally, Doug held the thick steel capsule in his hands. It had taken three missions, six weeks at sea across ten months, and $3 million to get this far.

Doug had to open the canister immediately, which made him unusually nervous. He worked with black boxes all the time, but always under the highly controlled conditions of his lab in Washington, DC. Normally, he'd take his time, carefully documenting his progress. But he didn't have that luxury. A team from the VDR's manufacturer in Sweden was waiting to hear that all was well—that the unit had been successfully recovered from three miles down and that it was in decent condition, not waterlogged or corroded. As soon as they got that message from Doug, they'd board a plane for the United States and meet him at NTSB headquarters to oversee the downloading of whatever data their device had managed to capture.

All people aboard the *Apache* crowded around Doug, anxious to see what was inside the canister—whether it had resisted the ocean's forces for so long—but he wanted to work under the best conditions possible. Changes in pressure from fifteen thousand feet below to the water's surface could have created an explosive condition inside the capsule.

Doug retreated to a supply room where he'd prepared a makeshift lab, carrying a cooler filled with seawater in which the VDR sloshed around. A few men from the team followed to watch Doug and shoot photos.

Wearing safety glasses, black latex gloves, and his navy blue T-shirt emblazoned with a yellow NTSB on the back, Doug lifted the capsule out of its watery bed and placed it on a table, protected with a white towel. The outer steel cylinder was about fourteen inches high and eight inches in diameter with a screw-on top, sealed with rubber gaskets. Remnants of the reflective tape that

Eric Stolzenberg had seen on the second mission still stuck to the canister, though a lot had worn off during the VDR's journey, leaving traces of rough adhesive ringing the can. Two steel handles welded onto one end had made it easier for the CURVE's pincers to grasp the thing three miles down, and Doug could see fine marks left by the robot's metal appendages.

Doug used hand tools—a couple of socket wrenches, adjustable pliers, and Allen wrenches—to deconstruct the base of the outer canister and pry it open. Claylike insulating material packed inside to protect the inner canister spilled out, leaving a pile of brown dirt on his towel. A plasterlike polymer—designed to withstand the great pressure of a deep ocean—surrounded the inner steel cylinder. Doug delicately chipped away at it like an archaeologist, keeping the pieces for future study.

The inner steel canister was four and a half inches long, one and a half inches in diameter. Doug clamped the head of it in a vise and slowly unscrewed the top.

A circuit board the size of half a graham cracker slid out. This was where all the information was stored, on one of those microchips. He breathed a sigh of relief. After spending so much time under intense pressure, the inner chamber was dry.

When Doug got the VDR back to DC in August 2016, the Swedish representatives of the unit's manufacturer met with him to help connect the memory chip to a forensic write-blocker to prevent accidental erasing of the media while downloading it. They made a copy of the information so that Doug could work off a facsimile, not the original, again preventing any possibility of error. And then the whole assembly got tossed into the red bin and filed away. It has sat there ever since. Officially, TOTE owns the VDR. If the company ever wants it back, it's theirs.

Doug wasn't sure what kind of data they'd be able to recover. He knew that sometimes things don't work out. Equipment fails all the time.

Like many twentysomething high-tech specialists, Doug has a matter-of-fact attitude about his work. He's passionate about the scientific challenges of extricating data from mangled iPads and steel boxes, but he isn't emotionally affected by what he might eventually hear. He's listened to pilots' final words before fatal collisions. He's heard train engineers in the midst of cataclysmic crashes. He's heard copilots and crew catching their first and last look at the plane, bird, or mountain that will kill them.

When I asked him what it was like to listen to *El Faro*'s audio, he said blankly, "It's my job," and shrugged, knowing he couldn't give me what I was looking for, silently wondering why reporters keep asking him this. What's there to say?

Doug loaded the manufacturer's playback software and visually scanned the digital representation of the audio. Twenty-six hours of something had been caught on the chip, but would it be clear enough to transcribe? Would it be white noise or gibberish? Or worse, had the VDR stopped recording two years ago?

Did they even have a recording of the right voyage?

To find out, he put on his headphones and skipped to the end of the tape. There was lots of noise from the wind outside, things crashing. And there was yelling, panic. He adjusted his acoustic filters and listened harder. The sounds of a ship's final moments were unmistakable. Doug had no doubt they had caught *El Faro*'s demise.

Twenty-six hours of audio on a black box leading up to a fatal accident was unprecedented for the NTSB. Usually they got thirty minutes, maybe two hours max. This particular VDR model, designed for the marine environment, was supposed to record a minimum of twelve hours. It's a looped system that writes over as time continues, and the chip was big enough to hold more time, so it simply kept recording.

Doug exported the five audio tracks from the six microphones installed in the ceiling of *El Faro*'s bridge into separate files. Each track would be heard through software that had various filtering capabilities. He spent the next two days quickly listening to the entire VDR, making notes, creating charts, orienting himself to milestones, like watch changes.

Most accidents aren't as dramatic and protracted as *El Faro*. For plane accidents, it's usually a quick hit. Doug says that the pilot might say "mountain" or "crane" and then hit it and the audio goes dead. But this tragedy unfolded over the course of days. Doug knew that, at the very least, they would transcribe every word of the last four hours.

The NTSB and US Coast Guard assembled a small expert team to listen to and type up what they thought they heard. Each member of the team sat in the listening room wearing a pair of headphones. Doug would play a snippet of the audio, and they would tell him what they thought *El Faro*'s crew said, and who they thought said it. Danielle was easy to distinguish from the others because she was a woman. But others sounded very similar to one another. If there was a disagreement, there would be a discussion. They wanted to get everything just right. No one wanted to misquote the dead.

They started with the last four hours leading up to the sinking.

The end of *El Faro* audio was very difficult to hear, not just because it captured a heartbreaking moment, but because as the ship sailed closer to the hurricane's eye, the quality of the audio signal significantly degraded. There was more wind howling around the navigation bridge, and the boat was shaking more, creating a lot of background noise. The blackout curtain around the chart room hung from ball bearings that slid in a metal track; they constantly rattled when the boat shook, making conversations there more difficult to hear.

By the final hour, all the audio channels except one were blown out; Doug was left with two mics—one over the chart table, one over bridge left—feeding into a single channel. If a layperson were just to play it as recorded, they wouldn't be able to hear much of anything. They'd hear a lot of hissing and some voices in the distance. Fortunately, as the winds picked up, the people on the bridge spoke louder, which helped the team make out their words.

Then Doug went back to the beginning of the audio, before dawn on September 30 when Davidson and Shultz were on the bridge. That conversation was easy to transcribe.

The watch change at midnight at the chart table proved very challenging, obscured by the blackout curtain rattling in its metal track.

At times, Doug had to play the same three words over a hundred times before everyone listening could agree on what was being said. He was constantly putting the audio through different filters to get a slightly different sound to help the team confirm what they thought they heard. They spent one day transcribing a total of nine minutes of audio.

In the twenty-six hours of recording, there were hours and hours of silence while the two people on the bridge—officer and helmsman—drove the ship and watched for hazards, steeped in their own thoughts.

It took twenty-two days, working ten hours a day, for the team to transcribe the VDR. The result was a five-hundred-page transcript, longer than any ever produced by the NTSB.

Doug may not have been affected by the audio, but the ex-mariners on the team were. The voices of those lost haunted them and turned a tragic case into a deeply personal one. The dead whispered in their ears long after they took off their headphones.

CHAPTER 31

TWENTY-FOUR MINUTES

At seven o'clock on the morning of October 1, Rich Pusatere was trying desperately to restart *El Faro*'s propeller. He lacked enough lube oil to overcome *El Faro*'s list, so the giant turbines, gears, and driveshaft sat frozen in time where they'd lurched to a halt an hour before.

Without power to steer her through the waves, the ship surrendered. She rolled onto her side like a wounded whale, easing deeper into the water, while thirty-foot surges ravenously tore at her decks, rocking her hard, filling her holds with the weight of the sea.

A dozen stories above him, Shultz said, "I think that water level's rising, Captain."

"Do you know where it's coming from?" Davidson replied.

At that point, Shultz had finally secured the open hatch cover above three-hold and had witnessed the mountain of water building up there. "I buttoned that scuttle up as hard as I could, and I don't know what else I can do," he told Davidson.

While down there, Shultz also had a conversation with Pusatere who told him that when they got thrown over to port, a car may have careened into the fire main, rupturing its connection to the sea. Water might have rushed in through that hole. When they started pumping water using the fire suppression system, he guessed, they were inadvertently pumping floodwater from one hold into another, redistributing the weight.

"When you went down there before, was there anything near the fire main?" Davidson asked Shultz.

"I couldn't see because the water level's too high," the chief mate replied. "The fire main was below the water, dark black water. And I saw cars bobbing around."

"The cars are floating in three-hold," Davidson said. Then he added with a laugh, "That makes them submarines."

"My concern is stability," said Shultz. "I have no concept of how much water may be sitting down there."

Davidson used the phone to call his chief engineer. "You think this list is getting worse?" Pusatere must have said yes. "Yeah," Davidson said. "Me too."

Davidson decided that pumping out water was their only hope, but they didn't know which compartments were flooded. In desperation, he told his engineer to open all the valves on the manifold so that they could pump all holds at once. But if any of those holds were water-free, air would get into the system and destroy the prime, rendering it useless. He was willing to take that risk.

Then the engine room called up: the bilge alarm in two-alpha hold went off. *El Faro*'s watertight doors between cargo holds were beginning to fail. Water was working its way from giant room to giant room, distributing liquid weight evenly along the inside of port hull, pulling her farther down. They were losing buoyancy.

Jeff Mathias called up to the bridge and informed Davidson

that the situation was dire. Then he asked about the downflooding angle of *El Faro*.

Davidson didn't know about the ship's downflooding angle. He wasn't familiar with the term.

"What's it called again?" Davidson asked Mathias over the phone. "Okay, we'll check that."

The downflooding angle of *El Faro* could be found somewhere in the chief engineer's office, Mathias told him. So Davidson asked Pusatere to look for it. But before he did that, he told Pusatere that he was going to ring the general alarm and wake everybody up. "We're definitely not in good shape right now. Just trying to control that list and see where the water's coming from."

He asked Shultz to make a round on second deck and see what he could see. His chief mate gave him a stricken look. The winds were up to at least 130 miles per hour. No one could stand on the second deck without getting swept right out to sea.

"You all right?" Davidson asked him.

"Yeah," Shultz said. "But I'm not sure I wanna go on second deck. I'll open a door down there and look out. It's chest-deep water washing over the deck."

"That's fine," Davidson said.

At 7:27, the captain told his chief mate and chief engineer that he was going to ring the general alarm. He wanted everyone to muster on the starboard side, the high side of the ship. "Make sure everybody has their immersion suits and stand by. Get a good head count."

Containers were peeling away and crashing into the water. The noise of the wind and waves on the bridge was deafening. Two minutes later, Davidson ordered abandon ship. "Tell 'em we're goin' in!" he yelled.

Danielle ran down to her cabin to grab her life vest, the captain's, and one for Frank. She would not return. Captain Michael Davidson and Frank Hamm were left alone on the bridge.

"Okay, buddy, relax," Davidson told his helmsman.

He looked out over the deck of his sinking ship into the howling dawn. In the whiteout of the foam, spray, and sea, he could see stacks of containers breaking free from their chains and yawning over the side before plunging down into the depths. "Bow is down," he observed. "Bow is down."

Over the radio, Davidson told his crew to throw their rafts in the water and get off the ship. But how could they even walk out onto the deck in those winds, let alone deploy a life raft? Everything—people, rafts, life suits—would be whipped away by Joaquin and into the waves, or thrown back against the ship's steel hull to be crushed. The air was solid with salt and water. You couldn't breathe out there. The crew probably crowded around the door leading to the second deck watching in horror the hellish world beyond through a porthole. Their survival instincts kept them there, huddled together.

El Faro rolled farther into the wind, exhausted by the fight, until her deck edge dipped into the brine. Superheated Caribbean waters beckoned her in. The ship's floors turned to the sky and became walls, her walls became ceilings. She was going gently into the eternal night of the deep ocean.

Two people remained on the bridge as she sank.

"Captain," Frank Hamm pleaded. "Captain. Captain."

Davidson braced himself on the high side of the bridge, looking down what was now the steep ramp of the floor. At the end of it, the heavy seaman was pinned to the corner by gravity and fear. He couldn't climb up to the starboard side of the bridge to get out. The angle of the floor was too steep.

"Come on, Frank," Davidson said. "We gotta move. You gotta get up. You gotta snap out of it. And we gotta get out."

"Okay."

"Come up."

"Help me."

"You gotta get to safety, Frank."

"Help me," Frank cried out.

"Frank, don't panic. Don't panic. Work your way up here."

"I can't."

"You're okay. Come on, don't freeze up, Frank. Come on."

"Where are the life preservers on the bridge?" Davidson shouted into the din. No one answered. He turned back to Frank. "Follow me."

"I can't."

"Yes, you can."

"My feet are slipping. I'm going down."

"You're not going down," Davidson yelled.

"I need a ladder."

"We don't have a ladder. I don't have a line."

"You're gonna leave me."

"I'm not leaving you. Let's go."

"I need someone to help me."

"I'm the only one here."

"I can't. I can't. I'm a goner."

"No, you're not."

"Just help me."

"Frank, let's go," Captain Davidson said. "It's time to come this way."

CHAPTER 32

SPIRITS

In the summer of 2015, Jill Jackson-d'Entremont packed up her house in Bucks County, outside of Philadelphia, and moved to Florida. She missed living near her childhood home of New Orleans and missed her two bachelor brothers, especially Jack Jackson. He'd spent so much of his life shipping out that she rarely got to spend much time with him. Now that the siblings were close to retirement age, Jill wanted to encourage her brother to focus on his art—his painting and sculpture. Jack had talent.

By late September, Jill was in her new apartment, surrounded by moving boxes. She hadn't set up her internet yet, but being disconnected for a few days was fine by her. With a presidential election coming up, she didn't need to know everything going on with the world. It was nice to be unplugged for a while.

Jill had taken charge of her life and should have been happy, but as September drew to a close, she began to have feelings of dread. What was it? The stress of relocation was finally behind her, so it couldn't be that. Maybe it was her health? She felt jittery.

On October 1 at 7:45 in the morning, Jill thought she was having a heart attack. A heavy pressure bore on her heart and she felt waves of panic shudder through her body. She resisted the urge to go to the hospital because she didn't want to mess with her insurance company, but she felt like she was dying.

Later that evening when she was visiting with friends, Jill found out about her brother's ship. The disappearance of *El Faro* was all over the news. Her friends said *Jack's a tough guy. He'll be fine.* Jill considered going to Jacksonville to join the rest of the families awaiting news, but her friends encouraged her to stay. They said *this could go one of two ways: either he's alive, and they'll visit him in the hospital and he'll get better because he's a strong guy, or he's gone.* So she didn't go. She waited at home. There wasn't anything she could do anyway.

Jill tells me that what she felt that morning was Jack's plasma leaving his body.

Her brother Glen was listed as Jack's emergency contact, but he never got TOTE's call. He lived in New Orleans and his landline was down—a car had hit a telephone pole a few days before the accident. He didn't have internet access, either; he stubbornly clung to his archaic flip phone. Like his sister, he didn't feel he needed to be fully plugged into the news cycle.

After *El Faro* sank, Jill moved to an old house with three fireplaces in New Orleans, close to Glen. She kept a photo of Jack on her mantel and talked to the photo by electric candlelight while eating dinner every night. When she was packing for the Jacksonville hearings, she put Jack's photo in her bag. That night, after washing dishes, a teacup jumped from the drying rack into the sink. Her first thought was that the house moved. A day later, it happened again. She decided Jack didn't like being smothered in her bag. She took out his photo and returned it to the mantel. The next night, the cups stayed put.

Jenn Mathias didn't believe in the supernatural. Whenever

she went to church, she'd always think to herself, *Yeah, right*. She only wanted to talk about things she can see or hear. But after her husband, Jeff, died, she went to a medium. She wanted to find out whether he'd suffered during his final moments.

On the morning before I met Jenn for the first time, I walked my daughter to school and observed a flock of cardinals. They usually travel in pairs—a male and female—and I thought the ten or so in the tree above us was unusual enough that I stopped and stared. "Isn't that weird?" I said to my daughter.

A few hours later, I was in Jenn's home enjoying a cup of coffee when she mentioned that she'd visited a medium. "What did he say?" I asked.

"He told me that Jeff communicates through birds. And Jeff's bird is a cardinal."

Weeks after Rochelle Hamm's husband died, she got a phone call from a stranger in southern Florida. They'd found her husband's helmet—it had washed up on the beach. She showed me the photos on her phone of the green plastic helmet. It looked brand-new, no scratches whatsoever, clearly labeled "Frank" with a black Sharpie. Nothing else of *El Faro* washed ashore. She says her husband sent it to back to her, to let her know that he loved her.

Marlena Porter says that her husband's last words to her were, *Meet me in the Bahamas*. At the time, she didn't know why he said that. But now she does. *El Faro* went down off the coast of the Bahamas. That's where her husband will always be.

"God allowed it," Pastor Robert Green declares of the *El Faro* tragedy. The pastor, who lost his stepson, LaShawn, shares the mysticism he finds in the names and numbers of this horrendous event. In Spanish, *el faro* means lighthouse—both a warning of dangers and a light in the darkness. The lighthouse has been used as a metaphor for God—a beacon of hope—for centuries. Even the storm's name, Joaquin, is the Spanish version of the Hebrew name Joachim, which translates to "raised by God."

Thirty-three people died when *El Faro* sailed into the storm and to Pastor Green, that's significant. "Thirty-three means something," he says emphatically. In the Judeo-Christian tradition, thirty-three is the number of promise, the numerical equivalent of amen. The name of God, Elohim, appears thirty-three times in Genesis. God demanded that Noah build an ark dozens of times; his final request, number 33, came with the covenant that he would never again flood the world. In the New Testament, Jesus lived thirty-three years and in that time accomplished thirty-three miracles.

The number 33 in Kabbalah signifies "the end of suffering." In Islam, thirty-three is al-Azim, the supreme glory, and considered the "perfect age"—how old believers will always be in heaven. The Tibetan Book of the Dead describes thirty-three heavens ruled by Indra and thirty-three ruled by Mara. And consider science: the human spine has thirty-three vertebrae, and human DNA is made up of thirty-three turns.

"God had a reason for sending this ship down," Pastor Green concludes. "Shipping is an industry that needs to be led, guided. These seamen are so invisible that nobody's concerned about their lives when they're shopping in Walmart. We don't think about the containers tied by hands to the decks. But these seamen are part of our society—navigating perils to bring us cars, gas, and other things. God wants to give us a warning. Shawn was one person but his life will touch one hundred thousand people."

In Maine, a week after Danielle was lost, her cat, Spot, suddenly passed away. He was young and healthy and was accustomed to long stints without her. At Danielle's memorial service, the clouds above formed the unmistakable shape of an anchor.

After the final family meeting in Florida, when relatives were told that the coast guard was ending the search, Deb Roberts and her husband navigated Florida's rush-hour traffic to get to the beach, any beach. Eight years before, Deb had driven to Castine to see her son Michael Holland off before he sailed to Europe aboard the Maine Maritime Academy's training ship. Now she was saying a different good-bye.

Holding her shoes, she walked to the water to be closer to her son. When she finally felt the ocean lapping at her legs, she says, "I had this beautiful moment on the beach where I just felt Michael. I was crying, really crying hard. I was leaning down, sobbing, and then this huge wave came and got me soaking wet. I knew it was Michael. And I was, like, *All right, Michael, I get it, I get it. OK, I'll stop.*"

One fall day a year after the accident, I visited Frank and Lillian Pusatere's home north of New York City on the Hudson. They were lovely, warm, and generous. We talked about their son Rich, their unfathomable loss, and what could have gone wrong on the ship. They took me to Rich's childhood room. It was small and neat with a world map pinned above the twin bed marked with places Rich had traveled on the ships. As I studied the map, Lillian casually mentioned that three-year-old Josephine still talks to her daddy sometimes. At the end of every conversation, Rich reminds his daughter that he will always love her.

EPILOGUE

El Faro was created in one age and destroyed in another.

In the four decades since she was built, Earth's oceans have absorbed an astounding amount of heat—20 x 10,000,000,000,000,000,000,000 joules—equivalent to 360 times the total amount of energy used by all people on Earth in a year.

We know that this man-made heat is penetrating the seas more than a mile down, creating a deep, rich supply of warm water to power increasingly devastating hurricanes.

It's payback time.

The 2017 Atlantic hurricane season generated the highest total accumulated cyclone energy ever recorded. Multiple Category 5 storms tore up the Caribbean islands and coastal United States. In September and October of that year, historic downpours in Houston and Florida (Hurricane Harvey), unfettered destruction in the Virgin Islands (Hurricane Irma), and total devastation in Puerto Rico (Hurricane Maria) left millions of people without electricity, potable water, homes, and access to medical care.

During that single season, hurricanes caused more than $188 billion in damages and took dozens of lives.

That money could have been spent building renewable energy infrastructure for our children.

Hurricane Joaquin was the strongest October hurricane to hit the Bahamas since 1866, and the strongest Atlantic hurricane of nontropical origin in the satellite era. Its recorded wind speeds hit 155 miles per hour. Its lowest pressure was 931 millibars, close to a record.

At eight o'clock on the morning on October 1, an Air National Guard Hurricane Hunter aircraft determined that Hurricane Joaquin's center hit 942 millibars, proving that *El Faro* had drawn remarkably close to the eye. At that point, the hurricane's 120-mile-per-hour sustained winds whipped thirty-five miles around its center.

Yes, we have the science to predict hurricanes, but that science is only as good as the models we use to understand our world and the data we feed them. On land, weather stations are everywhere, recording minute changes in humidity, temperature, and pressure. At sea, there is significantly less atmospheric data available, and even spottier information about ocean temperatures and the depths of those warm wells—all critical to hurricane modeling.

Thanks to improved technology, however, meteorologists have gotten much better at predicting a hurricane's track. In the early 1970s, one seventy-two-hour NHC forecast projected a hurricane's path nearly seven hundred miles off its actual course. Even now, however, the further out the forecast, the higher the error rate. In 2015, the seventy-two-hour track error averages run below one hundred miles, but one 120-hour forecast that year was nearly 350 miles off base, an error large enough to stick out like a big middle finger on the NHC's neatly descending error rate chart. That hurricane was Joaquin.

Meteorologists' ability to forecast how a hurricane will intensify hasn't improved significantly in forty years. The NHC still can't accurately predict whether a storm will whip itself into a frenzy or fizzle out. Nearly all its models, including its flagship—the Global Forecast System—told meteorologists that Joaquin would succumb to shear forces and quickly disperse.

But one model nailed both Joaquin's track and intensity from the beginning. It was the European Centre for Medium-Range Weather Forecasts, or ECMWF. As the dissenting voice among the more than fifty models that the NHC forecaster had at his fingertips, the ECMWF was easy to dismiss.

It's an open secret in the meteorological community that the ECMWF is consistently better than the NWS. The one question that gnaws at America's scientists is why.

The ECMWF is Europe's equivalent to the National Weather Service, but it draws funding and talent from more than thirty countries. Its 350-plus multilingual, multicultural staff not only gathers data and predicts weather, it also offers its findings to all of Europe, including Iceland and Turkey. The center actively engages with university researchers to enhance climate understanding. Unlike the cash-strapped National Weather Service, the ECMWF, with an annual budget of $100 million, offers robust workshops and research grants to draw a large scientific community to help improve its capabilities.

Much of the ECMWF's efforts go to refining its formidable computer model, which runs on some of the world's largest supercomputers—absolutely necessary to crunch the inordinate amount of data required to create an accurate picture of the complex and dynamic global weather forces. (It's important to note that the European center's model relies on a wealth of free climate data from the United States, but charges high prices to non-E.U. entities for its forecasting packages.) All of this computing power has made the model more accurate than ever.

It should go without saying that the researchers working at the

European center embrace science and are supported by an international community that values evidence-based conclusions. Make no mistake—no one spends a minute there debating whether climate change is real.

Few American politicians understand the importance of NOAA's work—the parent agency of the NWS. Politicians see it as an easily cut line item. After all, who are its constituents? A few scientists with their satellites, Hurricane Hunters, and weather balloons seem like a colossal waste of money. What's more, their data—which presents incontestable proof that oceans and the atmosphere are warming at alarming rates—is an affront to climate-change deniers.

Perhaps, then, it should come as no surprise that a May 2017 White House budget proposed to cut NOAA's budget by 16 percent. The *Washington Post* reported: "The budget 'blue book' for NOAA, which details the administration's funding recommendations, specifically directs the agency to 'reduce investment in numerical weather prediction modeling.' It calls for a $5 million funding cut 'to slow the transition of advanced modeling research into operations for improved warnings and forecasts.'"

Antonio Busalacchi, president of the University Corporation for Atmospheric Research, said the cuts "would have serious repercussions for the US economy and national security, and for the ability to protect life and property."

Without advanced weather prediction tools, Americans will have less accurate information about imminent hurricanes, tornadoes, and flooding—catastrophic events that will continue to escalate as climate change gains speed. Shipping and air travel will become riskier propositions, as all industries will have less time to prepare or avoid increasingly intense storm systems. Every citizen will shoulder the financial burden when insurance companies demand higher premiums to cover widespread damages. Costs everywhere, for everything, will grow dramatically.

Another proposed White House budget cut: 14 percent, or $86 million, from the US Coast Guard. The agency spends the bulk of its budget on stopping illegal drugs from entering the United States. In 2016, the coast guard seized $6 billion worth of cocaine and seven thousand people trying to illegally enter the country. (Notably, marijuana seizures have dropped significantly since the drug became legal in many US states.)

But the pride of the organization is its elite rescue swimmer program, which has saved an incalculable number of lives since it was launched in the mid-1980s, including 33,500 people in the aftermath of Hurricane Katrina. These heroic saves depend on top-notch equipment and training.

Less known, but perhaps critically important, is the coast guard's role as a regulator of America's commercial vessels. It's the kind of work that doesn't make headlines, until there's a colossal tragedy and everyone looks for someone to blame.

Several high-profile critics, most notably Vice Admiral James Card, have questioned the coast guard's expanded role under the Department of Homeland Security. In a damning report issued in 2007, Card wrote, "The Coast Guard has had a long and proud tradition of serving the country and marine industry through a robust and very professional Marine Safety program. U.S. safety standards, U.S. inspections, and the U.S. licensing system have been models for the rest of the world.

"Many point to the Coast Guard's increased role in Maritime Security and its move to the Department of Homeland Security as primary reasons for the deterioration [of its safety programs]."

The sinking of *El Faro* triggered much soul-searching at coast guard headquarters in Washington, DC. Captain Neubauer was devastated by the loss, he told me a year later; he took the sinking personally. "Before the accident, I didn't think a US vessel would undergo something like that," he said. "Something of that size. We had the inspections program in place and I thought that with all the modern technology available, we'd never have something like the *Marine Electric* again. I thought we'd done the job. I thought all those big, unsafe vessels had been taken out of the fleet."

In late 2017, the NTSB and US Coast Guard each released a report summarizing their findings of the *El Faro* investigation. Families of those lost, mariners, and the media consumed these reports, seeking answers to the whys and hows of this great tragedy.

The coast guard report provided a detailed timeline of *El Faro*'s final voyage, much of it based on analysis of the VDR, expert testimony, and study of *El Yunque*. Ultimately, the report lay the lion's share of blame on Captain Davidson. At the press conference, Neubauer admitted that Davidson would have been stripped of his license, based on the evidence they had. He had made several fatal mistakes. He ignored his colleagues' warnings and stayed in his cabin throughout the night when he should have been up on the bridge. He failed to practice the most basic weather avoidance procedures and deliberately put his ship and crew in the gravest danger.

Mariners' responses to the coast guard's assessment were mixed. Those who work at sea are superstitious and do not relish speaking ill of the dead, especially to landlubbers. They want to protect their own. And while they do acknowledge that Davidson's judgment was off, they want us to understand that he didn't act alone.

From the ill-fated design of *El Faro* to TOTE's leadership vacuum to lack of government oversight, professional mariners know that it takes legions of bad decisions and judgment errors to sink a ship.

At the very least, these mariners had hoped that the coast guard would shine a light on TOTE's failings as a company. Few of TOTE's shoreside executives were qualified for their jobs, and some may have been grossly unqualified. They thought that the crewing and human resources managers made a mess of a delicate staffing situation and left their officers and crew divided and distracted; people were promoted and fired without proper evaluation. But as long as the ships were reaching ports at appointed times and making money, TOTE's executives seemed satisfied.

The official accident reports addressed TOTE's restructuring and cronyism only glancingly. TOTE received a few service violations for minor issues that added up to a slap on the wrist.

TOTE's lawyers, paid for by the shipper's insurance company, rushed heartbroken and desperate family members to settle. A tempting payout—half a million dollars—would be theirs if they gave up the fight and agreed never to sue the company again. After months of waiting for their percentage of these comparatively modest settlements, the families' lawyers encouraged their clients to take the money. They wouldn't have their day in court. One by one, families reluctantly accepted the money and went home to pick up the pieces of their lives.

The cargo insurers, however, relentlessly searched for ways to get their money back. In the spring of 2017, they launched a $7 million lawsuit against StormGeo, the Norwegian company that owns BVS. Lawyers representing the cargo insurers argued that the BVS software provided "outdated and erroneous" data about the hurricane's path, which led Davidson to his fateful decisions. "The late captain was clearly unaware of the delayed and inaccurate hurricane locations and projections being proffered by the BVS 7," the complaint said. "Tragically, so strongly did the late captain trust the accuracy of the BVS 7 product that, when, on several occasions the vessel's mates suggested a change of course, he rejected those suggestions."

TOTE was unrepentant to the end. The company didn't even bother to file an internal incident and investigation record for the sinking of *El Faro*. At TOTE headquarters, it was as if the accident never happened.

In 2016, TOTE sent *El Yunque* to Alaska as a backup ship in the Pacific Northwest trade. She steamed through the Panama Canal and up the California coast, and finally docked in Tacoma, Washington. This time, US Coast Guard inspectors knew what they were looking for. They boarded her carrying the small steel hammers that they use to check the condition of steel on ships. When they tapped at parts of *El Yunque*'s ventilation trunks, their hammers went right through them. The steel had turned to dust.

The coast guard demanded that TOTE rebuild the trunks, but the company decided that making their ship safe would cost too much money. *El Yunque* made her last voyage from Tacoma to Brownsville. Empty, covered in rust, she limped to Texas and was scrapped.

Captain Neubauer's group also issued thirty-one safety recommendations to the commandant, who would then decide which to pursue—and which to ignore. The first recommendation was requiring that all new and existing multihold ships have high-water alarms wired to the navigation bridge. Properly working alarms would have alerted Davidson to the flooding in his ship much earlier.

Second, the report suggested that the coast guard conduct a thorough review of regulations concerning ventilators and other hull openings and make sure that all these vulnerabilities are considered when calculating downflooding angles on vessels. Had Davidson known about his particular ship's vulnerabilities, he might have made different choices.

Third, eliminate open lifeboats for all oceangoing ships in the US commercial fleet. That should have happened long ago. It was one of the most important recommendations to emerge from the *Marine Electric* tragedy, but hulls built before 1986 had been grandfathered in. Why should a ship's age preclude it from safety equipment that's standard on all other vessels? In fact, shouldn't older ships be better equipped, since they're more vulnerable? If the US Coast Guard had had any power over the industry it regulated, it would have mandated enclosed lifeboats on *El Faro* when the ship was modified in 2003. That ruling might have saved lives. It's possible that Davidson delayed abandoning ship until the very end because he knew that *El Faro*'s open lifeboats would be useless in Joaquin. Sure enough, both lifeboats recovered after the ship went down—one floating, one on the ocean floor—had been severely damaged.

Most ships sailing under other flags have alarms when watertight closures, including scuttles, are breached. *El Faro* had none. The coast guard report recommended that all watertight doors and hatches set off alarms when they're opened. That would have instantly alerted the ship's officers that the three-hold scuttle on the second deck wasn't secured.

A network of CCTV cameras inside the holds would have alerted *El Faro*'s deck officers to what was happening in the holds without having to go down there. They would have been able to see cargo breaking loose, flooding, the source of that flooding. On *El Faro*, the mariners could only guess what was going on in the holds. The coast guard report recommended requiring cameras in unmanned spaces on all commercial vessels in the future.

The report also suggested that all VDRs on commercial ships be installed in a float-free arrangement. In the event of an emergency, the unit would float away from the vessel and trigger an emergency signal. This way, the VDR would not go down with the ship.

There was a sense, especially among the families of the unlicensed crew lost on *El Faro*, that the seamen lacked a way to communicate unsafe conditions anonymously to the coast guard. They may have feared retribution from TOTE if they complained. Rochelle Hamm in particular wanted to make sure that mariners at sea had access to a hotline to report critical safety concerns not being addressed by the ship's personnel. The recommendations in the report included a shipboard emergency alert system so that crew wouldn't have to ask officers permission to use the satellite phone. If the crew of *El Faro* had had access to one, maybe someone would have made a call to shore to alert TOTE or the coast guard that Davidson was pursuing a reckless course.

Stronger guidelines would have ensured that TOTE's safety management system adequately identified risks and contingency plans to protect its crew, and the report recommended that the coast guard do more than simply require that a company have a system in place. Instead, the organization should offer clear guidelines for what such a system should look like, and how it should be taught and implemented.

The coast guard report also called for a comprehensive evaluation of merchant mariner training and credentialing. It was clear in the transcript of the VDR that some of the bridge officers didn't have enough knowledge or experience to make a clear case for avoiding bad weather. Davidson did not understand the ship's stability and downflooding angles, and he could not effectively manage his crew during the emergency. Basic rules of thumb, such as Buys Ballot's Law, had been forgotten long ago. Had America's maritime academies failed the merchant marine?

The balance of the safety recommendations focused on the coast guard's oversight of third-party classification societies, such as ABS. Recommendation number thirty asked that the commandant consider creation of a Third Party Oversight National Center of Expertise to monitor all parties performing work on behalf of the US Coast Guard. This was another lesson learned in the *Marine Electric* investigation forty years ago, but legislators continue to cut funding to the coast guard, undermining its ability to regulate America's commercial fleet. The agency's only option is to entrust third parties with critical duties. One mariner at the hearings informed me that ABS was the "richest entity in the room."

Since the mid-1990s ABS has taken over about 90 percent of the inspection work of deep-draft vessels the coast guard once did. The $1 billion Houston-based nonprofit gets most of its revenue from the companies it's paid to monitor. Some say it's like the fox guarding the henhouse. After all, the shipping companies are its only client. They're paying the ABS to keep their crew safe, but their real interest is in keeping their insurance premiums low and keeping their vessels sailing. A ship in port is a ship that's losing money.

Many worry that ABS's conflicts of interests, fueled by corporate demands, conspire to make American shipping increasingly dangerous. Perhaps to attenuate the coast guard's regulatory zeal, the ABS recruits the agency's top-brass to leadership roles. Currently, Rear Admiral James Watson, formerly of the US Coast Guard, was ABS America's division president until Jamie Smith, another former coastie, took over in early 2017. Many move into the private sector after retiring from the coast guard; a high-level position at the ABS can pay more than $1 million a year in salaries and bonuses.

Others think of the ABS as a friendly, understanding, industry-focused organization. Which can only help America's shrinking commercial fleet. Instead of coasties with clipboards, former merchant seamen from the ABS survey the ships to ensure they're compliant with the latest safety standards. On short notice, they'll come any time of day or night when the vessel's in port. There's a rapport. It's not as contentious a relationship as merchant and coast guard.

The US Coast Guard randomly reviews a percentage of cases approved by the ABS and lately, they've been finding more and more errors—most recently, in 2014, 5 percent of cases revealed 38 percent discrepancies. In several cases, an ABS representative approved automation ill-suited for the job required of it. In other cases, the classification society said a vessel's design had structural integrity, but it actually wasn't in compliance with the latest standards. There have been problems with the use of fire protection materials not properly tested, increasing the risk of fire, which at sea can be catastrophic. The coast guard worries that the classification societies lack proper training, procedures, and processes in place to oversee America's aging fleet.

In fact, ABS failed *El Faro*. Like most American ships, *El Faro* was enrolled in the coast guard's Alternate Compliance Program, which shifted nearly all inspection responsibility from the coast guard to ABS. A full ship inspection is performed piecemeal over a period of five years by ABS's surveyors and contractors who come aboard specifically to inspect certain things on their list. They're supposed to keep an eye out for anything amiss, but they missed things.

El Faro's VDR, for example, was inspected by an external specialist on behalf of ABS. He overlooked the fact that the battery that powered *El Faro's* VDR's location pinger would expire in a few months and gave it a pass. By October 2015, it was dead.

Corporations may always fight for deregulation (euphemistically known as small government). And taxpayers will always pick up the bill. It's incredible to think that all the money the government and TOTE "saved" by cutting corners over the years was spent many times over following the accident—first to finance the massive search-and-rescue effort and then to cover the cost of the three VDR recovery missions, which alone totaled more than $3 million. Not to mention the countless hours the US Coast Guard and the NTSB devoted to investigating the sinking. And, most importantly, the deaths of thirty-three men and women—people who were only doing their jobs—and the immeasurable suffering of those who loved them.

It's the same twisted math that prevents us from investing in renewable infrastructure while committing billions of tax dollars to subsidize America's petroleum industry.

WHEN I WAS GROWING UP, I WAS TAUGHT THAT HISTORY WAS LINEAR. THAT we were on a great, sweeping march toward some grand and glorious future. I learned that our language, culture, and science continually build on the depth and breadth of acquired knowledge—our marvelous birthright.

The story of *El Faro* contradicts that linear narrative. In 2015, an American cargo ship ended up at the bottom of the sea, taking thirty-three people with her. It was the deadliest American maritime disaster since World War II. With all our sophisticated technology, how could an accident of this magnitude happen?

The tragic loss of *El Faro* and her crew serves as a dire warning against complacency.

Humankind may chart a noble course but progress, like every voyage, requires strong situational awareness and a vigilant helmsman.

CREW LIST

In memoriam of the *El Faro* 33

Louis Champa Jr. (Daytona Beach, Florida), Qualified Member of the Engine Department, 51

Roosevelt Clark (Jacksonville, Florida), Utility Person, 38

Sylvester Crawford Jr. (Lawrenceville, Georgia), 40

Michael Davidson (Windham, Maine), Captain, 53

Larry "Brookie" Davis (Jacksonville, Florida), Able Seaman, 63

Keith Griffin (Fort Myers, Florida), First Assistant Engineer, 33

Frank Hamm (Jacksonville, Florida), Able Seaman, 49

Joe Hargrove (Orange Park, Florida), Oiler, 65

Carey Hatch (Jacksonville, Florida), Able Seaman, 49

Michael Holland (North Wilton, Maine), Third Assistant Engineer, 25

Jack Jackson (Jacksonville, Florida), Able Seaman, 60

Jackie Jones (Jacksonville, Florida), Able Seaman, 38

Lonnie Jordan (Jacksonville, Florida), Messman, 35

Piotr Krause (Poland), Riding Crew, 27

Mitchell Kuflik (Brooklyn, New York), Third Assistant Engineer, 26

Roan Lightfoot (Jacksonville Beach, Florida), Bosun, 54

Jeffrey Mathias (Kingston, Massachusetts), Riding Crew Supervisor, 42

Dylan Meklin (Rockland, Maine), Third Assistant Engineer, 23

Marcin Nita (Poland), Riding Crew, 34

Jan Podgorski (Poland), Riding Crew, 43

James Porter (Jacksonville, Florida), Utility Person, 40

Richard Pusatere (Virginia Beach, Virginia), Chief Engineer, 34

Theodore Quammie (Jacksonville, Florida), Steward, 46

Danielle Randolph (Rockland, Maine), Second Mate, 34

Jeremie Riehm (Camden, Delaware), Third Mate, 46

LaShawn Rivera (Jacksonville, Florida), Chief Cook, 32

Howard Schoenly (Cape Coral, Florida), Second Assistant Engineer, 51

Steven Shultz (Roan Mountain, Tennessee), Chief Mate, 54

German Solar-Cortes (Orlando, Florida), Oiler, 51

Anthony Thomas (Jacksonville, Florida), Oiler, 47

Andrzej Truszkowski (Poland), Riding Crew, 51

Mariette Wright (St. Augustine, Florida/Brockton, MA), Utility Person, 51

Rafal Zdobych (Poland), Riding Crew, 42

ACKNOWLEDGMENTS

This book could not have been written without the help of countless people—from those who recorded the history of shipping over the centuries, to friends, mariners, and families directly affected by the loss of *El Faro*. Many of these people were named in the book, but some were not.

Paul Haley, a former chief mate of *El Morro*, told me hundreds of stories about the seaman's life. A native Mainer, Haley first signed up with the US Merchant Marine in the 1970s and spent the final sixteen years of his career at TOTE. His tales gave me insight into the plight of the American mariner while slowly revealing an industry in crisis. During Hurricane Joaquin, Haley was in constant contact with Second Mate Charlie Baird tracking *El Faro*. Since she was lost, he, like so many mariners around the world, carries a heaviness in his heart.

Every morning, the first email I read was the GCaptain newsletter, created by the innovative Captain John Konrad. An insightful compendium of global maritime news, GCaptain is necessary reading for people reporting on the industry. The online forum accompanying his news site provides a place for mariners to speak frankly about the problems of working at sea. It is a unifying force in a fractured field.

I cannot thank the US Coast Guard's team of media specialists enough for their assistance. Spokeswoman Alana Miller at DC headquarters was committed to transparency; she strived to make information about the Marine Board investigation accessible and made sure all of the Jacksonville hearings were available via Livestream so that family and friends could follow along. Her colleagues in Florida—Lieutenant Rachel Post and Lieutenant Commander Ryan Kelley—as well as Chief Warrant Officer Paul Roszkowski in Los Angeles at the Motion Picture and Television Liaison Office were equally generous with their time, support, and dedication to accurate reporting.

Peter Knudson, the public affairs officer at the National Transportation Safety Board, not only spent an entire day listening to me interview half a dozen NTSB investigators, he also saved me from the labyrinth of the federal government morass known as L'Enfant Plaza.

Maritime lawyer Chris Hug helped me navigate the complexities of admiralty law, and when he didn't have an answer for me, he always found someone who did, connecting me with Boston's small, tight-knit maritime attorney community, a remnant of a once-robust industry.

Deborah Moulton put her trust in me and encouraged Charlie Baird to open up about his experiences aboard *El Faro*. They both offered invaluable insights into the shipping world, and that made all the difference.

Mary Bryson welcomed me into her Jacksonville home while I was attending the hearings and accompanying her husband, Eric, on piloting excursions. I will never forget her kindness and her discriminating taste in cats.

It was a pleasure working with retired Navy and Army officer Michael Carr, a knowledgeable and sensitive truth-seeker who relentlessly analyzed the Marine Board hearings and shared his thoughtful conclusions with a wide range of journalists, mariners, and members of the military.

Author Robert Frump, who wrote the definitive story of the *Marine Electric* tragedy, generously offered his encouragement and insights. Bob is a true investigative journalist and his book, *Until the Sea Shall Free Them*, is required reading for anyone interested in the ongoing struggle between capitalism, regulation, and those who make their living on American ships.

To better understand life at sea, I took a Grimaldi car carrier/container ship from Italy to Baltimore in July 2017 with Captain Francesco Rago of the Italian merchant marine. A careful mariner and natural leader, Francesco revealed to this writer the subtle art of mastering a ship.

Mel Allen, editor of *Yankee* magazine, was an early believer in my ability to tell this story. I first pitched him the *El Faro* tragedy from the perspective of the Maine-based families of the lost mariners and Mel agreed to take a chance on me. Without his initial support, this book would not exist.

Boston magazine editor Chris Vogel gave me wide berth to investigate the *El Faro* story while I was working at the magazine. In the few years we worked together, I learned a tremendous amount about story structure from Chris.

I never really understood the adage "a friend in need" until I approached Hillary Rayport and Anupreeta Das, two incredibly smart women who willingly subjected themselves to a long slog through an unedited, deeply flawed rough draft of this book.

Many thanks to my agent, David Patterson, of Stuart Krichevsky Literary Agency, who took on this first-time author and found me the indomitable Denise Oswald, my fiercely dedicated editor at Ecco/HarperCollins. She was a tireless advocate of this project from the moment we met.

Finally, Sean Slade left me alone when I needed solitude and engaged when I needed his incomparable wisdom.

A NOTE ON SOURCES

In researching this book, I was fortunate to have access to a wealth of information made public by the NTSB and US Coast Guard during their investigation of the *El Faro* tragedy. After the disappearance of the ship, the NTSB and US Coast Guard jointly interviewed dozens of family members of the lost mariners, TOTE executives and employees, coast guard officers involved in the search and rescue effort, and maritime industry experts—conversations which were then transcribed and made public. The Marine Board of Investigation re-interviewed many of these witnesses, as well as additional experts and mariners, during the three two-week public hearings held in Jacksonville in 2016 and 2017, many of which I attended. When I couldn't be there in person, I followed the hearings via Livestream. All of these hearings were transcribed by the US Coast Guard and made public.

Later in 2017, the NTSB issued several chairman's reports summarizing the facts of the case as they understood them. These included research on meteorology, survivability, engineering, electronic data, naval architecture, and human factors. The NTSB also painstakingly transcribed the twenty-six hours of recordings on the VDR and uploaded that five-hundred-page document on its website. (The audio was not made public.) These reports, plus supporting documentation, proved invaluable to me. Both the NTSB and the coast guard issued final reports of their findings and recommendations at the end of 2017, available to the public.

In addition, I personally interviewed dozens of key witnesses, family members, TOTE mariners, members of the coast guard and NTSB, and maritime experts involved in the case, traveling from Maine to Washington, DC, to New Orleans to Florida.

I've met some ships, too. Over two frigid but memorable days in the spring of 2017, I interviewed former NTSB Investigator Tom Roth-Roffy on SUNY Maritime College's training ship *Empire State* (a steamship built in 1961) in Fort Schuyler, New York. Maine Maritime Academy was kind enough to allow me to tour their training ship *State of Maine* (a diesel ship launched in 1989), docked in Castine, Maine. I traveled with Commanders Mike Venturella and Mike Odom of the US Coast Guard to Philadelphia to tour SS *Wright* (formerly SS *Mormacsun*) a cargo steamship built in 1968, now owned by the American government and operated by Crowley Maritime. The car carrier/cargo ship *Grande Congo*, a diesel vessel built in 2010 for Grimaldi Group, was my home for twelve days in the summer of 2017.

Maritime history is well documented and several excellent books have been published on the subject. I am grateful for the impressive scholarship found in the following works: *The Way of the Ship: America's Maritime History Reenvisioned, 1600–2000*, by Alex Roland, W. Jeffrey Bolster, and Alexander Keyssar; *The Sea & Civilization: A Maritime History of the World* by Lincoln Paine; *The Maritime History of Massachusetts, 1783–1860* by Samuel Eliot Morison; *The Forgotten Heroes: The Heroic Story of the United States Merchant Marine* by Brian Herbert; *Until the Sea Shall Free Them: Life, Death and Survival in the Merchant Marine* by Robert Frump; *America and the Sea: A Maritime History* by Benjamin W. Labaree, William M. Fowler Jr., John B. Hattendorf, Jeffrey J. Safford, Edward W. Sloan, and Andrew W. German; and an obscure out of print volume published in 1958, *The Maritime Story: A Study in Labor-Management Relations* by Joseph P. Goldberg. And for anyone who still harbors a romantic view of the age of sail, *Two Years Before the Mast* by Richard Henry Dana Jr. will disabuse you of that notion.

INDEX